# Nexus Network Journal

## FROM MEDIAEVAL STONECUTTING TO PROJECTIVE GEOMETRY

José Calvo López, Guest Editor

VOLUME 13, NUMBER 3

Autumn  2011

KIM WILLIAMS BOOKS

Nexus Network Journal
Vol. 13
No. 3
Pp. 503–792
ISSN 1590-5896

## CONTENTS

José Calvo-López

School of Architecture and
Building Engineering
Technical University of
Cartagena
Ps. Alfonso XIII, 50
30203 Cartagena SPAIN
jose.calvo@upct.es

Keywords: Stonecutting,
Stereotomy, Descriptive
Geometry, Projective Geometry,
Gothic, Renaissance, Baroque,
Enlightenment, Villard de
Honnecourt, Mathes Roriczer,
Juan de Álava, Philibert De
l'Orme, Alonso de Vandelvira,
Ginés Martínez de Aranda,
Girard Desargues, Jules
Hardouin-Mansart, Jean-
Baptiste de la Rue, Amedée-
François Frézier, Gaspard
Monge, Jean-Victor Poncelet,
Jules Maillard de la Gournerie

Research

# From Mediaeval Stonecutting to Projective Geometry

**Abstract.** We tend to think about technology as the application of abstract science to practical problems, but sometimes the inverse is true, as in the case of modern orthogonal projections, which originated empirically in mediaeval workshops and only after a long historical process gave birth to abstract projective geometry. However, this evolution is marked by strong transformations in the media of knowledge transmission, the social groups that control these forms of knowledge, and the very nature of this branch of knowledge. This article charts these transformations, and also serves as an introduction to the eight articles in this special issue of the the NNJ which examime particular issues raised by these historical processes, such as rib vaults by Juan de Álava, the use of ovals at the Escorial, the surbased vault at Arles town hall, staircases in the treatise of Juan de Portor y Castro, axonometric drawing in stonecutting treatises, Frézier's treatise on stereotomy as an antecedent to Monge's Descriptive Geometry, Monge's studies on developable ruled surfaces, and Jules Maillard de la Gournerie's criticisms on Monge's system.

During the twentieth century, a great number of books and articles dealt with the history of linear perspective. In contrast, scholarly works about the historical development of orthographic projections [La Gournerie 1855; Loria 1921; Taton 1954; Sakarovitch 1997] are rather scarce. Such a state of affairs probably reflects an extended misconception; for many architects, engineers and mathematicians, orthogonal projection is a trivial operation, resulting simply from taking away a coordinate from a Cartesian system. However, from the standpoint of the history of science, such a conception is anachronistic and inconsistent with historical evidence; in fact, orthogonal projection appeared as the result of a complex process that lasted centuries and predates Descartes by more than four hundred years.

In particular, the use of orthogonal projection in Classical Antiquity and the Early Middle Ages is, at best, quite limited. The best known Roman plan, the *Forma Urbis Romae* does not involve projection [Sakarovitch 1997: 27-29, 34-36]; it is a bare foundation plan, obtained as a horizontal section of a vast ensemble of buildings; of course, the depiction of objects lying on the same plane cannot be taken as a proof of the use of any kind of projection. As for vertical projections, such as the full-scale working drawing for the pediment of the Pantheon, laid out in the outer terrace of the Mausoleum of Augustus, include only elements that are placed at close vertical planes, such as the cornices and the inner face of the pediment [Haselberger 1983; 1994]; this happens also in miniatures of the Early Middle Age depicting besieged cities or Heavenly Jerusalem. That is, orthographic representations of Antiquity and the Early Middle Ages cannot be considered projections in the architectural sense, since they represent at most a quite shallow portion of space.

Nexus Network Journal 13 (2011) 503–533        NEXUS NETWORK JOURNAL – VOL. 13, No. 3, 2011    **503**
DOI 10.1007/s00004-011-0081-5; *published online* 22 November 2011
© 2011 Kim Williams Books, Turin

When the draftsmen of Antiquity wanted to convey the impression of space, they resorted to other means of representation, in particular cavalier perspective and "transoblique" projections of the kind used much later by John Hedjuk [Eisenmann 1975]. We should be wary here of another frequent oversimplification. It is usual to remark that orthogonal projection involves no deformation, but this statement can be applied only to figures that are parallel to the projection plane. Actually, orthographic projections mask the figures that are orthogonal to projection planes, such as vertical walls in plans and side façades or flat rooftops in elevations, turning them into lines. Thus, if one wants to represent solids in space, one should resort to linear perspective, almost unknown in Antiquity, or a forerunner of axonometry.

Fig. 1. Ravenna, Mausoleum of Theodoric. Groin vault on the lower story, c. 520

In contrast, orthogonal projections of distant objects were used in the portfolio of Villard de Honnecourt [c. 1225; 2009] and a fair number of Gothic architectural drawings. Villard's plans for the cathedral of Meaux (fol. 15r) or the abbey at Vaucelles (fol. 17r) include both the section of the pillars and the horizontal projection of the vault ribs. Vertical projections of objects in distant planes are also present at Villard's well-known drawings for Reims cathedral (fol. 31v), including the exterior wall at aisle level and the high windows, both in the exterior and interior elevations. The emergence of this method of representation seems to be connected with such constructive practices as the use of the plumb line for the geometric control of Gothic vaults, attested much later by the manuscript of Rodrigo Gil de Hontañón [c. 1540, fols. 24v-25v]. Masons hung plumb lines from rib voussoirs to check that these pieces were lying over their theoretical positions, with the help of a full-size tracing drawn on planks laid on a scaffolding under the vault. The inclusion of different planes in elevations, deriving probably from the use of the square in stone dressing, does not seem to advance so quickly [Rabasa 2007]; such elevations as the ones for the cathedral of Strasbourg, the baptistery of Siena or the one in the "Reims palimpsest" are not much bolder that the tracings for Greek and Roman pediments. By contrast, such plans as the one for a pinnacle in Vienna involve a plethora of horizontal projections of the different stages of the pinnacle, superimposed on one another [Recht 1995, 38-43, 59-66; Branner 1958; Ackerman 1997; Sanabria 1984, 78-82].

In any case, the full development of such methods took more than two centuries. Late mediaeval plans and elevations depart frequently from present-day rules of orthogonal projection; for example, it is fairly usual to represent roses in oblique planes as perfect circles, when they should be depicted as ellipses. During this process, and probably as a result of their efforts to overcome such difficulties, draftsmen or masons found that horizontal and vertical projections can be coordinated in some way [Sanabria 1984: 60-82]. An early fourteenth-century elevation of a bell-tower, probably copied from the Giotto project for the *campanile* at Florence cathedral, as a model for that of the Siena cathedral [Recht 1995: 63-67], includes a precisely constructed octagonal upper story, topped by a spire. Such a drawing cannot be prepared without starting from a plan or resorting to complex calculations regarding the geometry of the octagon, which were quite probably beyond the reach of masons and architects of the period. We must, therefore, surmise that masons or architects in fourteenth-century Tuscany mastered a method for constructing elevations starting from the plan.

Fig. 2. Laon, Cathedral. Sexpartite vault over the nave, c. 1180

However, the implementation of such procedures seems to have been neither quick nor easy. A century later, Mathes Roriczer's *Buchlein der fialen gerechtigkeit* [1486] explains how to obtain the elevation of a pinnacle starting from the plan; rather than using reference lines, as any student of Descriptive Geometry would have done, he draws an axis of the elevation, constructs orthogonals to this axis, and painstakingly transfers measures from the plan to these perpendicular lines in the elevation. Once again, such a procedure bears the mark of stonecutters' full-size tracings, inscribed on floors or walls or drawn on planks. In such tracings, it is not easy to construct parallels, while in contrast, orthogonals can be drawn with relative ease with a large set square [Martínez de Aranda c. 1600: 16; Calvo-López 2000b: II, 82]; this explains the use of the axis of the pinnacle.

The late Middle Ages brought about another classical graphic tool: revolution. Late Gothic German books [Bucher 1972; Recht 1995: 111] or Renaissance texts such as those by Hernán Ruiz [c. 1550], De l'Orme [1567] or Vandelvira [c. 1585] compute the spatial position of keystones in rib vaults through an idiosyncratic procedure. While ribs that are parallel to the vertical projection plane project as circular arcs, oblique ribs should project as ellipses. Masons of the period had no clear understanding of the ellipse; instead they used four-centre ovals for surbased arches, simplifying the tracing procedure, as we shall see below when we discuss Ana López Mozo's paper. Besides, an oval or an ellipse would have been useless for the dressing of the rib voussoir; by contrast, a true-scale representation of the axis of the ribs, obtained by revolution around a vertical axis, is quite useful. In particular, German masters carried this process to levels of high virtuosity in the *Prinzipalbogen* technique, which involves multiple vertical rotations of a series of ribs in order to unfold them over a single vertical plane [Müller 1990; Tomlow 2009].

Fig. 3. Reims, Cathedral. Quadripartite vault over the nave, c. 1250

The introduction of the Renaissance in France and Spain posed new problems. Broadly speaking, Gothic construction involves a point and line paradigm, where ribs and their intersections at the keystones provide a supporting network for loosely defined severies. In contrast, Renaisssance vault design stems from surface and volume: the faces of the voussoir are portions of the intrados surface and the bed joints, and must meet tightly with abutting voussoirs. New methods of geometrical control were required to build such architectural members; since vaulting in ashlar masonry is quite unusual in the Italian Renaissance, neither French and Spanish masons nor Italianate artists, who were responsible for much architectural work in the period, had previous experience with this problem.

French and Spanish masons, building on the Gothic tradition in some aspects, but departing from it in a number of essential traits, quickly put together a new set of geometrical tools to tackle these problems. Reference lines connecting plan and elevation appear twice in Dürer [1525/c. 2008: fol 15v, 84v]; in both passages the author mentions masonry lore. Forty years later, in the treatise of Philibert De l'Orme [1567] and the manuscripts of Alonso de Vandelvira [c. 1585] and Ginés Martínez de Aranda [c. 1600], reference lines are used everywhere, providing a method to correlate different projections that is clearly more efficient than Roriczer's axes. This allowed Renaissance masons to carry out easily transpositions of the vertical projection plane and to use orthogonal projections to control the dressing of the voussoirs of the most complex masonry members, generating projecting planes or cylinders by means of the square.

However, such procedure, known as squaring, is not the most efficient dressing method; in fact, it involves waste of both labour and material, as Philibert De l'Orme [1567, 73v] remarked. Thus, Renaissance masons devised a number of alternative methods, sometimes involving rotations around horizontal lines or graphical computations of the angles between voussoir edges and, quite frequently, the use of templates of the voussoir faces. These templates could be rigid, obtained by means of triangulations or rabattements, or flexible, based on a sophisticated system of cone developments that was known as early as 1543 [Palacios 1987; Palacios 1990: 18-20; Rabasa 1996b; Potié 1996; Ruiz de la Rosa 2002] and was fully mastered by De l'Orme's time [1567: 113]. Later on, Girard Desargues [1640; see also Bosse 1643] advocated the use of a suitably sloped plane instead of a horizontal projection plane, although his method was violently contested by traditional masons, led by Jacques Curabelle [1644].

Fig. 4. Amiens. Cathedral. Tierceron vault over the crossing, c. 1250

Such powerful graphical instruments were indispensable in Renaissance and Baroque ashlar construction, since the geometrical challenges posed by the architecture of the period were quite complex. Arches opened in oblique or sloping walls or at the junction of two walls generate elliptical openings; lunette vaults and arches in round walls bring about cylinder intersections; windows opened in domes involve cylinder and sphere intersections; splayed arches replace cylinders with cones, adding further complexity; rear-arches and stairs involve ruled surfaces (see, for example, [Jousse 1642] and [Derand 1643] and the articles in this issue by Giuseppe Fallacara et al., Rocío Carvajal and Snezana Lawrence].

All this led Jean-Baptiste De La Rue [1728], to include at the end of his *Traité de la coupe des pierres* a chapter on cylinder and cone sections, considered in abstract, under the name of *Petit traité de stéréotomie*. This last word, used for the first time by Curabelle in his *querelle* with Desargues, literally means "cutting of solids" and at first stood for the purely geometrical aspects of stonecutting. Ten years later, Amedée-François Frézier transformed De La Rue's short final appendix into a solid scientific foundation for the craft of masonry construction. As Marta Salvatore explains in her contribution to this issue, Frézier devoted the first volume of his *Théorie et pratique de la coupe des pierres et des bois ... ou traité de stéréotomie* [1737-1739] to stereotomy, a new science which encompassed projections, developments, angular measures and surface intersections, leaving the application of these concepts to actual stonecutting or *tomotechnie* for the other two volumes.

Salvatore also remarks that Frézier's *Stéréotomie* acted as a prodrome, or forerunner, to Gaspard Monge's Descriptive Geometry, a discipline that was intended to serve at the same time as a representational language for engineers and as a method for solving geometrical problems both abstract and applied. Monge, a former Professor of the Theory of Stonecutting at the Engineering School of Mézières, built his Descriptive Geometry on several foundations, including artillery and topography; however, double orthogonal projection, mastered by Monge thanks to his stonecutting teaching, played a central role in this new science [Monge 1798; La Gournerie 1855; Taton 1954; Sakarovitch 1992; Sakarovitch 1994a; Sakarovitch 1995; Sakarovitch 1997: 189-282].

Monge's approach poses a number of problems. First, the expression *Géométrie Descriptive* took over the semantic field that had been given to *Stéréotomie* by De La Rue and Frézier; as a result, the word "stereotomy" invaded the area that had been granted to *tomotechnie* and the meaning of the word – but only the *word* – passed quickly from theoretical and geometrical to practical and mechanic. In contrast, the discipline became progressively more abstract, dealing with such problems as the tangency between two surfaces in the *arrière-voussure de Marseille*, a rear-arch that rests on a round and a surbased arch. Nineteenth-century treatises solve this problem neatly using Hachette's theorem; however, as Sakarovitch [1992; 1997: 307-309] and Rabasa [1996a] have pointed out, it is impossible to distinguish "correct" rear-arches executed using this theorem from the ones built by masons in earlier periods, who smoothed the junction between the two surfaces empirically.

The separation of nineteenth-century Descriptive Geometry from actual constructive practices was shown clearly by the neglect of such practical methods as transpositions of projection planes, revolutions and rabattements. Monge had used them here and there in his lessons at the École Normale, but did not mention them explicitly. Later on, Olivier [1843-1844: in particular 18-19] campaigned for the inclusion of such procedures in the canon of Descriptive Geometry, dubbing them "the fundamental problems" of Descriptive Geometry. However, as Enrique Rabasa points out in his article in this issue, La Gournerie [1860: vi-viii; 1874: 154] remarked that changes of projection plane were not as new as Olivier hinted; at the same time, he made clear that the substitution of a sloping plane for the horizontal projection plane is useless in stone construction and other practical applications, since such an operation tampers with the natural position of horizontal and vertical planes.

Another important point that shows Monge's abstract approach is the issue of spatial representation. Gradually, writers on stonecutting had complemented their operational diagrams in double or multiple orthographic projection with more intuitive cavalier or linear perspectives, as Miguel Alonso Rodríguez, Elena Pliego de Andrés and Alberto Sanjurjo Álvarez explain in their contribution to this issue. Monge placed a heavy burden on the shoulders of orthographic projection: it had to fulfill both roles, operative and representative; in fact, he went as far as tearing cavalier perspectives out from the pages of the copy of De La Rue's treatise used at the École Polytechnique, as Rabasa explains in his article. However, orthographic projection is not the best method to represent volume, as I said at the beginning of this paper. This led Monge and the professors at the École Polytechnique to complement orthogonal projections with shades and shadows, building on similar practices at the École de Génie de Mézieres; this in turn led to the study of such elaborate problems as the shadow of the triangular-section screw [Sakarovitch 1997: 85-94].

While nineteen-century stereotomy languished in such trivial issues, Poncelet, a pupil of Monge at the École Polytechnique, was imprisoned in Saratov, in the wake of the Napoleonic Wars. Deprived of books, paper and pencil, he conceived Projective Geometry [1822], an abstract branch of mathematics dealing with the properties of figures that are left unchanged by projections. Thus, while closing its historical cycle, stonecutting lore furnished vital inputs for the creation of an important branch of mathematics; a common thread ties together mediaeval masons, Renaissance architects, enlightened engineers and nineteenth-century mathematicians, leading to the formation of an essential part of present-day science.

* * *

However, this evolution is punctuated with strong mutations. First, throughout the historical span that goes from the Late Middle Ages to the nineteenth century, this body of knowledge was snatched from the hands of one social group by another on several occasions. At the onset of the Renaissance in France and Spain, the control of the subject passed from masons, the heirs of the mediaeval tradition, to "professional architects", in Wilkinson's [1977] terminology. Of course, archetypal, "Albertian" Renaissance architects were not particularly interested in this field, since they focused on conception, rather than execution, and there was no strong tradition of vault construction in ashlar masonry in Italy. However, a new kind of architect emerged during the sixteenth century, in particular in France and Spain: a professional who had learnt architectural theory from Italian treatises, but who had also mastered constructive skills through hands-on practice, along the lines of the mediaeval tradition. Such figures as Philibert De l'Orme, put forward by Wilkinson as the archetype of this new kind of architect, or Hernán Ruiz and Alonso de Vandelvira in Spain, fit exactly into this model.

In any case, such a transition was not peaceful. Philibert De l'Orme remarks that the architect should master construction techniques in order to command the artisans, not the other way round [De l'Orme 1567: fol. 2r, 81r].

Fig. 5. Valencia, Blackfriars convent. Ribless tierceron vault over the Kings Chapel.
Françesc Baldomar, c. 1450

This is not empty rhetoric; at the *Salle de Bal* at Fontainebleau, he put in his proper place Gilles Le Breton, a mason who had previously had his way against Serlio. Anyhow, De l'Orme was seen as an intruder by both the masons of his period and by such humanists as Ronsard, who mocked him with a pungent epigram [Potié 1996; Pellegrino 1996]; a few decades later, Ginés Martínez de Aranda [c. 1600: iv] complained that "even if craftsmen swear they are knowledgeable in their trade, they are not granted the authority they deserve according to the labour of their studies". During the seventeenth century the field was taken over by the literate, in particular by clerics. With the exception of Mathurin Jousse [1642], a blacksmith, the main treatises of the period were written by the Augustine Fray Laurencio de San Nicolás [1639], the Jesuits François Derand [1643] and Claude-François Milliet de Challes [1674], the Cistercian Juan Caramuel y Lobkowitz [1678], the Theatine Guarino Guarini (published 1737, but written no later than 1683), the Oratorian Tomás Vicente Tosca [1707-1715], or by such a wealthy bourgeois and prominent geometer as Girard Desargues [1640].

After Tosca, the last significant example of clerical involvement in stereotomy, the torch was passed to military engineers, through the hands of a somewhat obscure transition figure, Jean-Baptiste De La Rue [1728]. Very little is known about him; Pérouse de Montclos [1982: 100] makes him an architect, taking into account that he submitted his treatise and a number of papers on quantity surveying to the Academie Royale d'Architecture; however, he also built bridges and designed machines for placing foundation poles. If he was really an architect, his *Petite traité de stéréotomie* and his links with the Academie suggest he was groomed in the scientific architectural tradition of François Blondel and Claude Perrault, both of whom were interested in stereotomy [see Swanson 2003; Gerbino 2002; Gerbino 2005]. Ten years later, Frézier, a military engineer in charge of the Brittany fortifications, took his cue with *La pratique de la coupe des pierres ... ou traité de stéréotomie* [1737-1739]; further on, Gaspard Monge made stereotomy a main subject of the École Polytechnique, as the main application of his *Descriptive Geometry* [1798].

From this moment on, the most significant examples of the huge pile of nineteen-century treatises on stereotomy were written by professors of the École Polytechnique [Sakarovitch 1994a]. It has been remarked that, after the Restoration the school of Monge turned into the school of Laplace, Poisson and Cauchy [Olivier 1852: xi-xii, xv-xviii; Sakarovitch 1997: 321-322]. In any case, this development was anticipated at the end of the clerics' period: both Milliet-Deschalles and Tosca include *lapidum sectione* and *cortes de cantería* as sections in their mathematical treatises, although it is well established that Tosca's treatise, at least, was widely used for the instruction of military engineers in Spain [Capel 1998; Calvo-López 2007].

Fig. 6. Peterborough, Cathedral. Fan vault over the retrochoir. Attr. to John Wastell, c. 1500

Along with these shifts in the identity of the social groups that control this field, the media of knowledge transmission changed accordingly. Carl Barnes [2001] has remarked that Villard's product is not an "Album", a blank book gradually filled with drawings, but rather a haphazard accumulation of sheets from Villard and other masters. As late as the eighteenth century, this approach is frequent in Spanish manuscripts, such as the one ascribed to Juan de Portor y Castro [1708], including sheets from different authors and subjects. Such a conception is still present in the first printed works in the field; it is still not clear if there are two or three booklets from Roriczer [1486; c. 1490], since they are frequently bound together. Later on, these manuals take the form of the Renaissance treatise; however, sixteenth- and early seventeenth-century books still adopt the didactical methods of the "how-to book", giving instructions for each particular piece without furnishing the proofs of the geometrical procedures involved. Such proofs gradually appear from the seventeenth century on, in particular in Frézier [1737-1739]; thus, the treatise evolves into the scientific monograph. Just before the extinction of the species in the nineteenth century, the last of these mutations took place; a heated debate about skew bridges was carried out in books, but also in journal articles [C.L.O. 1837; Spencer 1839; Barlow 1841; La Gournerie 1851; La Gournerie 1853; La Gournerie 1872; see also Becchi 2002].

The very nature of this branch of knowledge was affected by these social and formal shifts: starting as a purely empirical craft, it had taken on the striking form of an *experimental* geometry during the early seventeenth century. Shelby [1972] and Sanabria [1984] have stressed the fact that mediaeval masons' geometry is quite different from the practical geometry of Hugh of Saint Victor [c. 1125]; accordingly, they classify stonecutters' lore as "constructive geometry". Later on, Fray Laurencio de San Nicolás made an striking assertion: "by means of plaster models you will know that practical knowledge is in concordance with speculative knowledge, as I have experimented with my hands before writing" [1639: 70]. Thus, in Fray Laurencio's view, constructive geometry, in parallel with the learned geometry of the period, includes two branches, practical and speculative; in particular, the practical or experimental aspect of this branch of knowledge is carried out with the help of reduced-scale models. In fact, when Alonso de Vandelvira [c. 1585: 22r] is explaining a particularly difficult issue, the concavity or convexity of the templates for an arch opened in a round wall, he is afraid that the reader will be befuddled, and suggests that he prepare a model to verify his assertions empirically.

Fig. 7. Annaberg, Church of St. Anne. Net vault over the nave, c. 1500

However, a significant break took place at this moment. The clash between Desargues and Curabelle led to the proposal of a singular kind of duel: two groups of masons, each under the direction of one of the contenders, were to build an oblique arch according to the systems of their leaders; a substantial reward of two hundred *pistoles* was to be given to the winning team. However, Desargues pointed out that the correctness of an arch should not be judged by masons, but rather by geometricians; in other words, for Desargues, abstract geometry lies on a superior plane than empirical validation by means of practical stonecutting. In the end, the duel did not take place, since Curabelle obviously took the opposite stance [Desargues 1648; Sakarovitch 1994b]. This episode signals the final blow to empirical geometry; in our days, experimentation is mandatory in physics and mathematics, but is strictly forbidden in geometry. However, in the field of stereotomy, the resistance to abstract rationalism lasted for a long period; as Rabasa points out in his paper for this issue, La Gournerie [1874] mocked Olivier [1843-1844], saying that if he had read Frézier [1737-1739] he would have known that the transposition of the horizontal projection plane, employed by Desargues [1640; see also Sakarovitch 2009b], is useless in stereotomy and had been rejected by practitioners.

Thus, the long thread that leads from mediaeval stonecutting to projective geometry poses a fair number of open problems for interdisciplinary research, including, for example, orthogonal projections in Gothic architectural drawings and their limitations; the transformations of orthogonal projection at the transition between Gothic and Renaissance; the multiple exchanges between masons, architects, engineers and cartographers in the Early Modern period; the distinctive traits of French treatises as opposed to Spanish manuscripts; the confrontation between empiricism and rationalism in Early Modern stonecutting, exemplified by the Desargues-Curabelle duel; the transition from stonecutting, artillery and other sciences to Descriptive Geometry during the Enlightenment and the French Revolution; and the historical roots of Projective Geometry in stereotomy and other practical disciplines.

In consequence, a Call for Papers was made for a monographic session at the eighth Nexus Conference on Relationships between Architecture and Mathematics, held in Porto, Portugal, 13-15 June 2010, asking for studies on the formal, social and epistemological shifts in stonecutting and stereotomy, in orthogonal projections and architectural drawing, and in descriptive and projective geometry. Of course, such a wide array of topics cannot be covered by a single conference session or even an entire issue of a journal; still, four papers were read in Porto, while other four papers have been selected for inclusion in this issue, focusing on a number of key problems in these subjects.

*     *     *

To start with, in "Geometric Tools in Juan de Álava's Stonecutting Workshop" José Carlos Palacios and Rafael Martín deal with ribbed vaults in the school of Juan de Álava, a Spanish mason of the early sixteenth century, clearly influenced by late German Gothic, noted for his impeccable continuity between pillars and ribs. After briefly outlining Juan de Álava's career, the author puts forward a clear explanation of the origin and evolution of rib vaults. Groin vaults were frequently used in Roman architecture. They involve two horizontal, orthogonal cylinders of equal radii; the intersection of both cylinders results in two ellipses, the groins that give its name to this vault type. During the Imperial period, such vaults were often built in concrete and therefore did not require ribs. In the High Middle Ages, when commerce in the Mediterranean was risky and pozzolana was in short supply, such vaults were sometimes built in ashlar masonry, solving the union of the cylinders by means of L-shaped voussoirs that cross the intersection line and belong to both cylinders. However, such a solution, which was used, for example, in the lower story of the sixth-century Mausoleum of Theodoric in Ravenna (fig. 1), requires careful control of the dressing process and was not widely used in the Middle Ages; in fact, it was eschewed in the upper story of the Mausoleum for the well-known monolithic vault.

In the Romanesque period the groin vault was still used, in particular in church aisles. However, when building in rubble and mortar, it is not easy to obtain a neat profile at the groins, while the use of L-shaped voussoirs in ashlar masonry was forgotten, as we have seen. Thus, Romanesque builders began using ribs placed at the groins, to mask the junction between both cylinders, either in rubble or hewn masonry. At the same time, masons understood that, since ribs must be placed before the body of the vault, the ensemble could be built dispensing with the heavy formwork and centering required to build a ribless groin vault; in fact, a rib vault can be constructed using substantial centering for the ribs, and only a slight centering for the sections between ribs, known as severies, dispensing at the same time with real formwork.

Fig. 8. Santiago de Compostela, Cathedral. Horizontal ridge rib vault over the cloister galleries. Juan de Álava, 1521-1527

In any case, the construction of the elliptical ribs required to follow the groin profile was quite difficult, since the curvature of an elliptical arc changes constantly. In constructive terms, this requires the use of a different template for each voussoir of the rib. Early Gothic masons radically simplified this problem by raising the profile of the vault and using round diagonal arches at the groins. This required raising the transverse arches as well; however, instead of using raised elliptical arches, builders resorted to the pointed arch for wall and transverse arches. In addition to solving this problem, pointed arches allow for a remarkable simplification and rationalisation of the building process, resembling twentieth-century prefabrication, since all voussoirs of a pointed arch are identical, except the keystone, which can be materialised by a pair of obliquely cut voussoirs. Pushing this idea to the limit, pointed arches of different spans, and even diagonal round arches, can be built using only one kind of voussoir, as a diagram in Villard's portfolio, attributed to the anonymous draftsman known as Hand IV, suggests [Villard de Honnecourt c. 1225: 21v; 2009: 13, 150; Branner 1963; Shelby 1969; Bechmann 1991].

Later, from the mid-thirteenth century on, masons multiplied the ribs, starting with a pair of tiercerons and a lierne at the Amiens crossing (fig. 4); gradually, such networks of ribs grew in complexity, including double tiercerons, ridge ribs and many kinds of straight or curved liernes. Along with its decorative function, such a network of ribs makes it possible to place the stones in the severies directly over the ribs, dispensing with any kind of centering for the panels. However, such complexity required new methods of geometric control. In a square or rectangular quadripartite vault, involving only diagonal ribs and wall or transverse arches, the symmetry of the vault guarantees that both diagonal ribs meet naturally in space. By contrast, in complex Late-Gothic vaulting, a series of sophisticated procedures were required to assure that ribs meet precisely at the keystones, both in plan and elevation.

The main section of Palacios and Martín's contribution to this issue focuses on these problems. Taking as a case study a number of vaults built by Juan de Álava in the cathedrals of Santiago de Compostela and Plasencia, the Blackfriars convent of San Esteban in Salamanca, and the San Marcos priory of the military order of Santiago in León, they examine the methods used by Álava to control the spatial layout of the vaults while keeping the radius of all ribs equal where possible. In particular, they classify Álava's vaults in four types. First, they deal with *rampante llano* or almost horizontal ridge rib vaults, such as the ones at the cloister of the cathedral of Santiago de Compostela (fig. 8). After this, they analyse the nave vault in San Esteban and the vault in the presbytery of Plasencia cathedral, which use horizontal longitudinal ridge ribs, while transverse ridge ribs feature a remarkable slope; the resulting effect resembles a pointed barrel vault. Next, they study the vaults in the cloister of San Esteban, featuring strictly horizontal ridge ribs, akin to the ones in English Gothic.

Palacios and Martín finish their typology with a peculiarly Spanish kind of vault. While in French or English churches it is customary to place the choir at the head of the church, behind the crossing, in Spanish and Southern French cathedrals it is usually placed before the crossing. In small churches with small or no aisles, this location is impractical and the choir is placed usually at the foot of the church, in a mezzanine placed over a surbased vault. The authors deal with a pair of such vaults, located at the *sotocoros*, or under-choirs, of San Marcos in León and San Esteban in Salamanca, remarking that the use of three-centre arches allows Álava to build extremely surbased vaults through the use of powerful geometrical methods.

*   *   *

Ana López Mozo's paper, "Ovals for Any Given Proportion in Architecture: A Layout Possibly Known in the Sixteenth Century," tackles a significant problem regarding the empirical methods of builders and architects during the Renaissance. As Palacios and Martín stress in their contribution, the three-centre basket-handle arch was used widely along the Late Gothic period. To construct the directrix of such arch is relatively easy; besides, since it involves arcs with only two different radii, it is quite convenient for practical stonecutting: the voussoirs for such an arch can be dressed using only two kinds of false squares with a curved arm, known as *biveaux* or *baiveles*. However, to adjust a three-centre arch to a given span and rise is not so easy. Renaissance masons mastered a number of methods for the tracing of four-centre ovals, such as the four well-known solutions explained by Serlio and other similar procedures included in the manuscripts of Hernán Ruiz and Vandelvira; of course, these methods can be applied both to the layout of oval vaults and the tracing of basket-handle arches [Serlio 1545: 17v; Ruiz c. 1550: 24v; Vandelvira c. 1585: 18r; Huerta 2007]. Each of these procedures is only valid for a given proportion between span and rise. With the onset of the Renaissance, such restrictions began to be felt as unacceptable; the need for a tracing method that made it possible to construct a surbased arch of a given span and height was seen as an urgent issue. The problem was tackled frequently through the use of an affine transformation of the circumference. Although masonry literature furnishes a number of different solutions to this problem, stonecutters essentially started by tracing a circumference whose radius was coincident with the rise of half the span of the arch they were intending to build, and adapted it to the other dimension by means of the affine transformation. Although such constructions furnish points of an ellipse, masons were usually not interested in such abstract concept; in fact, they usually joined these points by means of arcs of a circle passing through three consecutive points [Dürer 1525; Serlio 1545: 13v-15r; L'Orme 1561: 12r-13r; Ruiz c. 1550: 37r-37v; Vandelvira c. 1585: 18v; Martínez de Aranda c. 1600: 1-2].

Fig. 9. Anet. Chateâu, Hemispherical dome over the chapel. Philibert De l'Orme, c. 1550

During the intensive research on the vaults of the Escorial for her Ph.D dissertation [2009], López Mozo found a number of basket-handle arches that do not fit in the solutions furnished by Serlio, Hernán Ruiz and Vandelvira for three-centre half-ovals. After performing a thorough survey of oval-tracing methods in French and Spanish stonecutting literature, she remarks that no procedure for the tracing of an oval of a given proportion is documented in stonecutting literature before the treatise of Tomás Vicente Tosca [1707-1715; the section on stonecutting was published in 1712]. However, according to López Mozo, it is not realistic to suppose that the Escorial basket-handle arches were traced as half-ellipses. Along with the practical considerations about *baiveles* and voussoirs, she remarks that arches such as those in the Basilica narthex include a moulding that is parallel to the edge of the arch, and there is no practical solution for tracing a curve that maintains a constant distance from an ellipse. Thus, the presence in the Escorial of ovals with different proportions, in particular in the main church narthex, leads her to posit the hypothesis of a general tracing for three-center arches of any proportion between span and rise, used by the stonecutters at the Escorial. Such a method is perfectly possible in theory, and is in fact included in Tosca's treatise, although there is no direct evidence of its use in the Renaissance.

At the same time López Mozo points out the presence of a drawing about conic sections in the correspondence of Juan de Herrera, the architect of the Escorial in its later phase; besides, the presence in his personal library of Federigo Commandino's translation of Apollonius's *Book on Conics* [Apollonius Pergaei 1566] shows he was interested in the theory of the conic sections. All this leads us to an interesting issue. Although such abstract mathematical concepts as conic sections and ellipses are apparently of no interest to stonecutters, there are some exceptions. First, Dürer [1525/c. 2008: 16r-18r] furnishes a sensible procedure for the construction of conic sections, thirty years before the publication of Commandino's translation, based on the use of orthographic projections; quite significantly, he mentions explicitly that he is using stonecutters' methods, as I remarked earlier. Although the method is exact in itself, the results are flawed; in particular, the ellipse is not symmetrical about two axes, and rather resembles an ovoid. Some years later, the manuscript of Hernán Ruiz reproduces Dürer's drawings, managing to present a more than acceptable result – especially considering the precision of the drawing instruments of the period – using reference lines, as in his stereotomic drawings.

That is, masons mastered a powerful geometrical tool – orthographic projection – which can be used to solve not only the practical problems of stonecutting, but also to tackle the abstract theorems of learned geometry. The two worlds, practical and theoretical, were beginning to make contact at this period, but still were separate realms; the proof of such state of events is that, as far as we know, neither Hernán Ruiz nor Herrera used the abstract geometry of conics for practical purposes, although it can be argued that both men were grappling for such solutions.

*     *     *

Fig. 10. Seville, Cathedral. Elliptical vault over the Chapter Hall.
Hernán Ruiz el Joven and Asensio de Maeda, c. 1560-1592

Giuseppe Fallacara, Fiore Resta, Nicoletta Spalucci and Luc Tamboréro have contributed a paper entitled "The Vault of the Hôtel de Ville of Arles" regarding an outstanding example of French stereotomy: the flat vault at the ground story at the Arles Town Hall (fig. 12).

Spanning an area 16 m x 16m with a rise of only 2.4 m, it includes two separate vaults, meeting at a hidden reinforcing arch; each of these vaults is penetrated by a number of lunettes. This remarkable achievement has always been seen as a masterpiece of the art of masonry. At the same time it poses an interesting case study of the power struggles between the corporations of artisans and the Parisian academy. After a number of construction failures and consultations with architects and masons, the direction of the execution was taken over from Dominique Pilleporte, mason, and Jacques Peytret, painter-architect, by Jules Hardouin-Mansart in 1673. However, Hardouin-Mansart had to leave for Béziers and he instructed for a month Peytret on the technical details of the vault construction, leaving him a number of drawings and templates. Thus, the local practitioners played the role of executors of the designs of the architect; although Peytret solved a number of important execution details and contributed a number of personal solutions, all in all he kept the general design of the vault furnished by Mansart, as Richard Etlin [2009] has shown recently.

Luc Tamboréro is a member of one of these corporations of ancient craftsmen, *Les Compagnons du Devoir*; he even uses such a traditional nickname as *La perseveránce d'Arles*. In order to finish their training period and be accepted as *compagnon fini*, the members of these corporations must present a model of an outstanding work, known as *chef-d'oeuvre*. Tamboréro fulfilled this requirement with a model of the Arles vault in stone, at 1:5 scale; this allowed him to study the stereotomic layout of the vault [Tamboréro 2003], stressing its connections with carpentry procedures. However, a number of questions remained to be answered.

Fig. 11. Paris, Abbaye de Val-de-Grace. Groin vaults with oval cross-section on the transverse vaults. François Mansart, Jacques Lemercier, Pierre le Muet and Gabriel Leduc, c. 1650

Tamboréro teamed up with a number of researchers from the Politecnico di Bari led by Giuseppe Fallacara, who have conducted a state-of-the-art survey of the vault, using a 3D laser scanner, presented in their contribution to this issue.

The detailed study of the vault carried out by the Bari team, together with Tamboréro's practical expertise, has allowed them to reach interesting conclusions. For example, they have compared the general dimensions of the ground floor room and the three façades of the Town Hall with traditional Provençal measurements in *cannes* and *pans*, while also hypothesizing that it may correspond to a composition based on the Golden Section. Second, they point out the idiosyncratic layout of some capitals of the columns along the walls, which are reminiscent of similar solutions by Juan Caramuel y Lobkowitz [1678]. Third, they stress the combined use of vertical plane curves, such as the intersections between the corner lunettes and the main vaults, and warped curves, at the junctions between the large vaults and the axial lunettes, as well as at the union between both main vaults.

This seems to indicate that the geometry of the Arles vault stems from two different traditions: the Gothic paradigm, where simple linear curves provide a network over which complex surfaces are laid out, which, according to Tamboréro [2003], reaches seventeenth-century masonry construction through carpentry; and the mainstream Renaissance system, carried on into the Baroque period, which starts from relatively simple surfaces that lead to complex non-planar intersections (see also [Calvo-López 2000a] for the alternating use of both systems on lunettes). A detailed image of the geometry of the vault, carried out starting from the data gathered with the scanner, together with the analysis of the archival documentation, leads Fallacara and his colleagues to hypothesize that the corner lunettes were set at the start of the tracing process; that is, planar curves were used at the beginning of the design and warped curves at the end.

Fig. 12. Arles. Town Hall. Surbased vault on the ground story.
Jules Hardouin-Mansart and Jacques Peytret, 1674

From this starting point, the authors analyse the issue of stonecutting methods used in the construction of the vault. First, they assert that the vault cannot have been dressed by squaring, since in this method there is no need for the detailed instructions given by Hardouin-Mansart to Peytret; besides, in this particular vault the "loss of stones" pointed out by De l'Orme would have been quite significant. Next, they discard also the possibility that the vault was made using templates, since all intrados surfaces are warped, and finally they hypothesize that it may have been done using bevels, known in French as *sauterelles*. Although there is no mention of bevels in the French stonecutting literature of the period (with the exception of Bosse, who was rejected by practising masons), the authors remark that the bevel is present in Diego López de Arenas' [1633] carpentry treatise; it is also quite frequent in Alonso de Vandelvira's stonecutting manuscript [c. 1585]. Of course, the centuries-old connections of Southern France with Spain [Pérouse de Montclos 1982: 200-212] support this theory.

*     *     *

Rocío Carvajal's paper, "Stairs in the Architecture Notebook of Juan de Portor y Castro: An Insight into Ruled Surfaces," deals with the last significant example of masons' personal notebooks, dated 1708 and prepared by Juan de Portor y Castro; this mason was apprenticed in Santiago de Compostela [Taín 1998, 67-68, 269] and also connected with another important stonecutting center, Granada. Many solutions in Portor's notebook recall the methods of Ginés Martínez de Aranda [c. 1600], and in fact Portor [1708: fol. 22; see also Gómez Martínez 1998: 38] mentions a contest for the post of master mason in Granada cathedral, won by Juan de Aranda Salazar, nephew and disciple of Martínez de Aranda.

Given the bulk of Portor's notebook, including more than one hundred stereotomic problems, Carvajal's contribution focuses on straight stairs. First, the author stresses masons' idiosyncratic methods of representation and analyses their evolution. Some drawings included in Vandelvira's manuscript may seem at first glance ordinary elevations; however, detailed inspection shows that all stair flights are shown in a side view, while according to present-day orthodox technical drawing conventions some flights should be shown in front view. Although Vandelvira's representation strategy may be considered highly unorthodox by current standards, it has clear benefits, allowing the mason to control easily the tracing of the stair and the dressing of its voussoirs. A century later, Portor follows an eclectic trail: he includes an orthodox elevation, constructed by means of reference lines drawn from the plan, but the real stonecutting procedure is based on a disassembled set of side views of the flights, just as Vandelvira's was.

Carvajal's paper focuses on a particular kind of stairs, those with straight-profile flights. Actual built examples of such stairs are quite scarce in Spain; the problem is usually solved by means of curved-profile flights, exemplified by the lost staircase of the Capitole at Toulouse and its Iberian precedents [Pérouse 1982: 168-169; Gómez-Ferrer 2009] and the stairs at the convent of Santa María de la Victoria and the Chancillería in Granada, mentioned by Vandelvira [c. 1585, 59r], which Portor must have known well. In contrast, straight-profile flights are much less usual, since they involve complex problems during execution. In fact, two of the most significant examples, the stair at the Casa de Contratación in Seville and the one at the convent of Santa Catalina in Talavera are considered important stereotomical archetypes, and are cited both by Fray Laurencio de San Nicolás [1639: 118v] and Portor [1708: fol. 15].

Carvajal explains that the intrados of such stair flights, in particular the Seville type, is a hyperbolic paraboloid, since it materialises a surface that passes through two straight lines that are neither parallel nor concurrent; of course the masons did not use such a name, nor did they think in terms of generatrices and directrixes. Portor explains, however, a number of different solutions to the same problem. These stairs can be drawn using longitudinal courses, drawn in parallel with the enclosing walls, which lead to transversal joints between the voussoirs in the same course; however, the problem can be solved the other way round, using transversal courses and longitudinal joints. As for curved profile stairs, transversal joints are usual in the earliest examples, in particular in the Catalonia and Valencia area, but reaching as far as Toulouse in the north and Lorca in the south; by contrast, longitudinal courses are typical of Andalusian examples, such as the Chancillería and Santa María de la Victoria. However, while Vandelvira focuses on longitudinal courses for curved profile stairs, in the Andalusian tradition, he solves straight-profile stairs with transversal courses, and in fact the most significant built examples of straight-profile stairs, the ones at Seville and Talavera, are solved with transversal joints.

In contrast, Portor tries to furnish both solutions for straight-profile stairs, longitudinal and transverse courses. In particular, in the transverse course solution he uses a hyperbolic paraboloid for the flight and a plane for the landing, trying to achieve continuity between the intrados surfaces of the stair. This poses a fairly sophisticated problem: the paraboloid can be laid out using a line in the plane as the edge generatrix, in order to achieve first level continuity between both surfaces; however, since the paraboloid is a warped or non-developable surface, the tangent plane at the edge generatrix is variable, while the tangent plane at the landing is of course the landing plane itself. In any case, as Carvajal remarks, masons were not hindered by such difficulty; they achieved continuity in a practical way, retouching at will the intrados surface along the junction between flight and landing.

Thus, this episode exemplifies an essential trait of stonecutters' approaches to geometrical problems: they used remarkably advanced notions, such as the hyperbolic paraboloid, were fully aware of the difference between developable and warped surfaces, and even tackled such complex problems as geometrical continuity, anticipating results that entered the realm of learned geometry only during the eighteenth century, as we shall see in the papers by Marta Salvatore and Snezana Lawrence. Anyhow, stonemasons' methods remained in the purely empirical realm, since their interest was not focused on abstract problems, but rather on practical execution.

*    *    *

At first glance, the paper entitled "Graphical Tools for an Epistemological Shift: The Contribution of Protoaxonometrical Drawing to the Development of Stonecutting Treatises," by Miguel Alonso, Elena Pliego and Alberto Sanjurjo, departs from the main thread of this issue, the connections between stereotomy and orthogonal projection, and focuses instead on the forerunners of axonometric drawings included in stonecutting treatises. However, the increasing presence of this particular kind of parallel projections in the treatises and manuscripts of De l'Orme [1567], Vandelvira [c. 1580], Martínez de Aranda [c. 1600], Derand [1643], De La Rue [1728] and Frézier [1737-1739] poses an uncomfortable question: since stonecutting is so closely tied with orthographic projection over the centuries from Villard [c. 1225] to Monge [1795, 1796], as we have seen, why are protoaxonometrics frequent in stonecutting literature? As the authors stress, the purpose of these representations is clearly didactic, in contrast with the strictly operative nature of orthographic diagrams in these treatises and manuscripts, which reproduce stonecutters' full-size tracings literally. In other words, since stonecutters' diagrams are devoid of any representative power, protoaxonometric drawings must supplement them, so that treatises can be understood by a wide, non-mason readership, as a result of an epistemological shift that was turning a closed craft practised by artisans into a branch of mathematics.

The lack of representative efficiency of the orthographic projections included in stonecutting treatises stems from two different issues. The first one is strictly related to masonry: inscribing a full-size tracing on a floor or a wall is slow and tiresome; thus, masons were prone to leave any unnecessary lines out of the tracing. The extreme economy of masonry tracings is mirrored in books and manuscripts; even such printed treatises as Cristóbal de Rojas's *Teoría y práctica de fortificación* [1598: 98v] dispense with arches' outer profiles. The other reason is more abstract and affects any kind of orthographic projection: as I remarked earlier, any plane that is orthogonal to the projection plane, such as wall faces in horizontal projections or floors and side façades in cross-sections, is depicted as a line. This geometrical fact disqualifies plans and elevations, as such, for volumetric representation. That notwithstanding, a number of complements to orthographic projections, such as shades, shadows and line weight, can lend these non-representations some illusion of volume.

After mentioning the Chinese and Renaissance precedents of axonometric drawing, Alonso, Pliego and Sanjurjo chart the evolution of protoaxonometrical representations in stonecutting literature from the first, scattered examples in De l'Orme [1567], Vandelvira [c. 1585] or Martínez de Aranda [c. 1600], to the frequent use of this kind of representation by Derand [1643] and the mastery of the technique by De La Rue [1728]. They stress such points as the early presence of worm's eye views in a scheme about voussoir dressing in Martínez de Aranda [c. 1600], applied afterwards to vault representation. The authors also explain the systematic procedure used by De La Rue to construct parallel projections, taking coordinates from plan and elevation and transferring them to cavalier perspectives. Such representations may be considered legitimate axonometrics, since De La Rue's technique implies the existence of an axis, although it is not explicitly depicted in the plate, in contrast to nineteenth-century drawing manuals.

The authors mention a particular detail that is quite significant for our purposes. De La Rue's treatise [1728] includes a curious feature in order to improve its didactic efficiency: here and there, some fold-out models in lightweight cardboard are included between the sheets of the book. At first sight, this might recall present-day children's pop-up books, but the intention of such fold-out models is not only didactic: one of them shows the errors in the solutions of De l'Orme [1567], Jousse [1642] and Derand [1643] for the sail vault over a square plan. Up to a certain extent, the empirical geometry of Vandelvira [c. 1585] and Curabelle [1644] is still alive here.

In any case, the detailed analysis of De La Rue's axonometric drawings and its didactic nature brings us back to the initial question: can orthographic projection be used as a representational vehicle to convey an intuitive depiction of volume and space? Probably, Gaspard Monge thought so: as mentioned, De La Rue's axonometric drawings were torn out of the copy of De La Rue's treatise used at the École Polytechnique. This violent reaction against two centuries of evolution towards the use of oblique parallel projection might have been provoked, again, by two different reasons.

Fig. 13. Paris, Church of Saint Sulpice. Flat vault with spiral joint.
Gilles-Marie Oppenord, 1714-1745

First, quite probably these protoaxonometries were considered unscientific by Monge's entourage. As far as I know, in Monge's period no one thought of cavalier or military perspective as projections of any kind (see [Rabasa 1999] for this problem in the nineteenth century). If Monge had conceived De La Rue's cavalier and military perspectives as oblique projections, he would quite probably have rejected them as arbitrary, in contrast to the canonical orthogonal projections. Second, following the tradition of the Mézières engineering school, Monge struggled to endow orthographic projections with the representative power they lack in themselves, through the use of shades and shadows [see Sakarovitch 1997: 81-89]. This explains the presence of this subject in a later edition of Monge's *Géométrie descriptive* edited by Brisson [Monge and Brisson 1820], based on Monge's lectures at the École Polytechnique.

*  *  *

As we have seen, the first volume of Frezier's three-volume *Pratique de la coupe des pierres et des bois ... ou traité de stéréotomie* [1737-1739], focuses on *stéréotomie*; during this period, the term stood for a branch of abstract geometry dealing with the division of solids. Only after thoroughly surveying this field in the first volume, Frézier deals with practical stonecutting or *tomotechnie* in the remaining two volumes. In particular, *stéréotomie* includes *tomomorphie*, the science of the form of sections, that is, the nature of curves resulting from the intersection of solids; *tomographie*, that is, the technique that makes it possible to draw these sections, either on a plane or on a surface; *ichnographie* and *ortographie*, which deal with the orthogonal projections of solids, either on horizontal or vertical planes; *épipedographie*, the science of surface developments; and *goniographie*, which encompasses methods for computing the real size of angles depicted in orthogonal projection.

Thus, Frézier turned the small, ancillary abstract appendix at the end of De La Rue's treatise into a formidable scientific basis for the craft of stonecutting. At the same time, he laid the foundations for the old and new science of Descriptive Geometry, as Marta Salvatore stresses in her article, "Prodromes of Descriptive Geometry in the *Traité de stéréotomie* by Amédée Francois Frézier". In particular, Frézier combined the graphical methods of previous stonecutting treatises, such as those by De l'Orme [1567], Jousse [1642], Derand [1643] or De La Rue [1728], with the analytical work of such mathematicians as Alexis Claude Clairaut [1731], who had studied warped curves using multiple orthogonal projection. This allowed Clairaut to define these curves by reference to two or three planar curves, since any curve of double curvature can be seen as the intersection of two or more cylinders generated from planar directrixes; it is rather striking to find such an abstract concept explained as an operative stonecutting method in Frézier [1737-1739: II,13].

Taking this into account, Salvatore's contribution focuses on *tomomorphie*, that is, the properties of planar or warped curves resulting from the intersection of a pair of solids, in particular quadric surfaces. For example, given an elliptical cylinder, Frézier endeavours to find the plane that will give circular sections of the elliptical cylinder. In order to do so, he rotates the section plane along an axis of the elliptical directrix until it reaches a position where the second axis of the planar section equals the length of the first axis, which is fixed, since it is also acting as rotation axis. Such an operation is fairly easy in an elliptical cylinder, and in fact it finds precedents in stonecutting literature [Martínez de Aranda c. 1600, 16; Blondel 1673; Gerbino 2005]. However, the same procedure cannot be applied to an elliptical cone, since the centres of the circular sections do not lie on the axis of the cone; the solution to this problem would not be found until Theodore Olivier's period [1843-1844].

Another interesting issue in Frézier's *Tomomorphie* is found in his original contributions to the treatment of warped curves created by the intersection of quadric surfaces. While Clairaut applied analytical geometry to such curves, Frezier dealt with them in terms of synthetic geometry; such an approach was quite unusual at the time. Anyhow, this leads Frézier to explain in detail the nature of three different warped curves, which he dubs *cicloïmbre*, *ellipsïmbre* and *ellipsoidïmbre*, making use of an empirical metaphor: he suggests that the reader imagine a circle drawn in the spine of a book prepared for binding. If the bookbinder plays with the pages, sliding one over the other, the circle transforms itself into a warped curve, called *cicloïmbre* by Frézier, which is in fact the parallel projection of a circle upon a cylindrical surface; parallel projections of ellipses on quartic surfaces give *ellipsïmbres* as a result, while central projections of ellipses over cylinders furnish *ellipsoïmbres*.

Of course, Frézier did not neglect the application of geometry to actual stonecutting. In fact, he deals with practical matters in two-thirds of the treatise; however, in contrast to the empirical approach of older manuals, he always tackles construction problems from the standpoint of abstract geometry, sternly criticising the methods of previous authors, as Salvatore remarks. Thus, Desargues's somewhat flawed attempt is brought back to life: the abstract science of space takes precedence over the practical methods of stonecutters, which is understood rather as a technical application of pure science. However, Frézier's position is more balanced than that of Desargues; while Desargues simply does not take into account the work of De l'Orme, Frézier shows a thorough knowledge of his predecessors, if only to confute their solutions. Thus, Frézier's work stands at a crossroads: on the one hand, it is a rare example of a sound work on synthetic geometry in a century dominated by algebraic methods; on the other hand, he makes the decisive move in order to push construction procedures from the empirical field to the realm of rational geometry, tying a knot between abstract science and practical activity. All these factors make Frézier's treatise a forerunner of Monge's Descriptive Geometry, not only at the epistemological level, but also at the social one, since the sound application of science to the practices of craftsmen was to have the effect of shifting the status of such branch of knowledge upward, and potentially that of the craftsmen themselves, as Monge saw clearly.

\* \* \*

Snezana Lawrence has contributed a paper about a quite interesting subject: "Developable Surfaces: Their History and Application." When dealing with the connections between mathematics and the real, everyday world, we tend to think in terms of "applied science": abstract notions are conceived in the minds of scientists and are applied to practical issues only afterwards. The notion of ruled surface, as many other geometrical concepts, seems to arise from the inverse process; it appears in stonecutting treatises and manuscripts as early as the sixteenth century, but it only reaches the realm of learned mathematics around 1750, through the work of Monge [1769] and Euler [1772]. Euler's interest in surface developments may have been connected with cartography, and thus lies outside of the scope of our subject, but it is clear that Monge was attracted to the topic not only through topography, but also as a result of his experience as professor of stonecutting [Sakarovitch 1992].

Fit. 14. Paris, Church of Sainte Geneviève, now Panthéon. Lunettes, dome and oval vault.
Jean-Baptiste Rondelet and Maximilien Brébion, 1780-1790

Lawrence stresses in her article that Monge used developable surfaces to solve a problem that is connected both with artillery and topography: to compute the height a wall should have in order to protect a military position from enemy fire. In the mid-eighteenth century, the solution of this important issue involved long and complex calculations; Monge found a quick way to solve the problem on site by employing a plane that is tangent to the topographical surface and a developable tangential surface. Such a successful solution showcased Monge's outstanding mathematical capabilities and allowed him to rise within the ranks of the Military Engineering School at Mézières, breaking the social barriers due to by his humble origins. However, Monge was prohibited from publishing his discoveries, which were classified as military secrets; he could only explain his findings about developable surfaces in an abstract fashion, leaving out any possible military use of the concept.

As a result of his prestige in Mézières, Monge was granted the position of Professor of the Theory of Stonecutting. He must have felt a keen interest in developable surfaces and their application to stereotomy, since stonecutters had been involved with these surfaces for centuries, as we have seen, using either rigid or flexible templates to control the shapes of voussoir faces. The use of such instruments by masons shows a typical shift from the empirical to the analytical understanding of the problem. In earlier masonry literature, the developable surface is conceived as one where two finitely distant generatrices are parallel or concurrent; by contrast, in the modern notion of developable surface, the generatrices that need to be parallel or concurrent are placed at an infinitely small distance.

Of course, Monge assumes this last position, linking the issue with differential geometry. However, the roots of his interest in this subject lie in practical stonecutting, in particular in the construction of exactly fitting templates for intrados faces and bed joints, as Monge himself states in his *Descriptive Geometry*. This allowed him to manipulate such objects with remarkable spatial insight, obtaining powerful results about developable surfaces; in fact, Lawrence stresses that Monge's visual approach led him to superior results than the analytical methods used by Euler, who was blind by that time.

At the same time, Monge's abstract approach to the issue made him go to great lengths to confirm that the bed joints of an ellipsoidal vault with three different axes were developable surfaces. In order to do this, and at the same time keep this surfaces orthogonal to the ellipsoidal surface, he studied in depth the notion of curvature lines, which led him to adopt sloping joints for the ellipsoidal vault, departing from usual practice. Furthermore, he was enraptured by his discovery; he suggested that the roof of the main hall of the National Assembly should be constructed in the shape of an ellipsoidal vault with three different axes, featuring a network of joints along curvature lines. He stressed that the result would be as beautiful, but much less arbitrary, than the ribs on Gothic churches [Monge 1796]; in some way, he was trying to convert the room into a Temple of Reason, conceived in a mathematical way.

As Sakarovitch [1997: 309-313; 2009a] and Rabasa [2000: 296-302] have remarked, such an approach stretches the conventions of practical stonecutting beyond reasonable limits, since orthogonality between intrados and bed joints, on the one hand, and developable templates on the other, are advisable but not mandatory. In fact, no ellipsoidal vault with bed joints following curvature lines has ever been built, as far as I know. I will come back to the issue of Monge's abstract conceptions against practical procedures in the next section, when dealing with Rabasa's article about the opposition between Monge and La Gournerie, but for now it is worthwhile to recall a striking episode that marks the far end in the evolution of the subject from empirical constructive practice to abstract mathematical concepts: Lawrence points out that around 1900,

Lebesgue dealt with developable surfaces that are not ruled, although they cannot be used in the real world in which architecture takes place.

*    *    *

Last but certainly not least, in "La Gournerie versus Monge," Enrique Rabasa presents a significant episode that shows that the battle between mathematical abstraction and practical construction that had started with Desarges and Curabelle was still blazing in the mid-nineteen century. Olivier, a prominent heir of the Monge tradition, wanted to "enlarge the conquered realm" of Descriptive Geometry, in Gino Loria's [1921] words, and stressed the utility of such geometrical procedures as transpositions of the projection planes. Although present in Monge's original lessons in Descriptive Geometry, such changes were not stressed either by Monge or by his disciples, such as Hachette [1822] and Leroy [1834]. Olivier, in contrast, focused on them as the "fundamental problems" of Descriptive Geometry, using both transpositions of vertical and horizontal planes. Jules Maillard de La Gournerie, who succeeded Leroy at the École Polytechnique and Olivier in the Conservatoire National des Arts et Métiers, remarked that such methods were not new at all, since they were used by Desargues; he approved of changes of vertical projection plane, but remarked that substituting a sloping projection plane for horizontal one is useless in stereotomy, since there is no advantage in tampering with the natural position of the voussoirs inside the masonry. Thus, La Gournerie fired against Monge and Desargues by elevation, passing over Olivier's position; he stood against an abstract conception of geometry that makes the vertical and horizontal planes equal, neglecting the presence of gravity; that is, La Gournerie implicitly approved the secular lore of masons against the abstract conceptions of Monge's school.

Fig. 15. Dresden, Zwinger. Coffered vault, dome over pendentives, skew arch and octagonal vault on the ground access to the courtyard. Gottfried Semper, 1847-1855

Rabasa also emphasizes La Gournerie's reaction against Monge's abstract conceptions of such issues as axonometry and linear perspective, stressing that Monge and his assistants tore the drawings in cavalier perspective out of De La Rue's treatise, fearing that they would lure the students to an *arbitraire* method of spatial representation, which had a long and venerable history dating back to Classical antiquity. By contrast, La Gournerie shows a remarkable interest on axonometry, although he focuses on the abstract, isometric flavor of axonometry invented, for once, by an Englishman, Farish [1822], rather on the simplistic but effective cavalier perspective, which he dubs "rapid perspective", while also stressing that France was less advanced in this field than Britain and Germany.

The issue of linear perspective in La Gournerie is more complex. Monge had proposed as a perspective method the old projection-and-intersection method of Piero della Francesca [c. 1480]; although Monge [1820] explains the vanishing point method as an *abregé* or simplified procedure, it is quite clear that any departure from exact, abstract, central projection would be for him even more arbitrary and unscientific than axonometry. Against such a stance, La Gournerie starts by mocking the notion of an objective, scientific perspective, remarking that, in order to achieve an optimal trompe-l'oeil effect, perspectives should be seen closing one eye, from a particular point that is not marked at all. As a consequence, he underwrites all the licenses painters have taken along the centuries, drawing as circles spheres placed near the corners of the picture, which should be depicted as ellipses according to scientific central projection, or eschewing the fat columns that should be placed at the sides of the drawing, and so on. Although such stances may at first seem contradictory to his initial statement, it is crucial to understand that they are driven by skepticism, since La Gournerie rejects the concept of objective, rational perspective, and instead considers the result acceptable if pleasing to the eye of the observer; Curabelle would have wholeheartedly approved such a stance.

Another pungent issue addressed by Rabasa's article is the social extraction of both contenders. Monge was the son of a small merchant and could not hold any important post in the army of the Ancien Régime, which was open only to the nobility and, arbitrarily enough, to the sons of glass merchants. He climbed the ladder to the highest positions solely on the basis of personal scientific merit, but he must have suffered much scorn along the way. It is no wonder that when the Revolution came, he was a fervent Jacobin. By contrast, La Gournerie was the son of a nobleman who had fought with the Royalists at the Vendée and inherited from him the title of Viscount although, if we are to believe Laussedat [1897], he never used it along with his name. At first sight, a nobleman raising the flag of artisans' methods strikes us as a romantic, nostalgic, and somewhat contradictory figure. However, La Gournerie was not imprisoned in an ivory tower: like De l'Orme, Rojas and Frézier, he had been trained as an engineer in Brittany, working at the Ileaux de Bréhat lighthouse, placed on a rock that was only accessible during low tides. Thus, the key to La Gournerie's position seems to be skepticism, which fits his political conservatism well; like Principe Salina in Giuseppe Tomasi di Lampedusa's novel *Il Gattopardo,* he knew the Ancien Régime was lost forever, but he mistrusted the abstract rationalism of the Enlightenment, both in politics and in the "graphic arts".

### Acknowledgments

This article is included in the research project "Stonecutting technology in the Mediterranean and Atlantic areas. Survey and analysis of built examples" (BIA2009-14350-C02-02)", sponsored by the Ministry of Science and Innovation of the Spanish Government in the context of the National Plan for Research, Development and Innovation.

The author wishes to thank Kim Williams and João Pedro Xavier for the inspiring environment during the Porto edition of "Nexus 2010: Relationships Between Architecture and Mathematics", the authors of the papers included in this special issue for their penetrating insights, and María Dolores López for her help in the preparation of the manuscript.

All of the photographs that appear here are by the author.

## References

ACKERMAN, James S. 1997. Villard de Honnecourt's Drawings of Reims Cathedral: A Study in Architectural Representation. *Artibus et Historiae* **18**, 35: 41-49.

APOLLONIUS PERGAEI. 1566. *Apollonii Pergaei conicorum libri quattuor, una cum Pappi Alexandrini lemmatibus et commentariis Eutocii Ascalonitae. Sereni Antinsensis philosophi libri duo ... Quae omnia nuper Federicus Commandinus Urbinas... e graeco convertit et commentariis illustravit.* Bononiae: Ex Officina Alexandri Benatii.

BARLOW, W.H. 1841. On the skew bridges. *The Civil Engineer and Architect's Journal* **4**: 290-292.

BARNES, Carl F. Jr. 2001. What's in a Name? The Portfolio of Villard de Honnecourt. *AVISTA Forum* **12**, 2: 14-15.

BECCHI, Antonio and Federico FOCE. 2002. *Degli archi e delle volte. Arte del costruire tra meccanica e stereotomia.* Venice: Marsilio.

BECHMANN, Roland. 1991. *Villard de Honnecourt. La penseé technique au XIIIe siécle et sa communication.* Paris: Picard.

BLONDEL, François-Nicolas. 1673. *Résolution des quatre principaux problèmes d'architecture ...* Paris: Imprimerie Royale.

BOSSE, Abraham and Girard DESARGUES. 1643. *La practique du traict a preuues de M. Desargues ... pour la coupe des pierres en l'Architecture ...* Paris: Pierre des Hayes.

BRANNER, Robert. 1958. Drawings from a thirteenth-century architects' shop: the Reims palimpsest. *Journal of the Society of Architectural Historians* **47**, 1: 9-21.

———. 1963. Villard de Honnecourt, Reims and the origin of gothic architectural drawing. *Gazette des Beaux-Arts* **61**: 129-146.

BUCHER, François. 1972. The Dresden sketchbook of vault projection. Pp. 527-537 in *Proceedings of the 22nd Congress of Art History.* Budapest.

C.L.O. 1837-1838. On the construction of skew arches. *The Civil Engineer and Architect's Journal* 1: 279-280, 313-314.

CALVO-LÓPEZ, José. 2000a. Lunetas y arcos avanzados. El trazado de un elemento constructivo en los siglos XVI y XVII. Pp. 165-175 in *Actas del Tercer Congreso Nacional de Historia de la Construcción.* Madrid: Instituto Juan de Herrera.

———. 2000b. 'Cerramientos y trazas de montea' de Ginés Martínez de Aranda. Ph.D. thesis, Universidad Politécnica de Madrid.

———. 2007. Piezas singulares de cantería en la ingeniería y la arquitectura militar de Cartagena en el siglo XVIII. Pp. 167-176 in *Actas del Quinto Congreso Nacional de Historia de la Construcción.* Madrid: Instituto Juan de Herrera.

CAPEL, Horacio, Joan Eugeni SÁNCHEZ and Omar MONCADA. 1988. *De Palas a Minerva. La formación científica y la estructura institucional de los ingenieros militares en el siglo XVIII.* Madrid.

CARAMUEL Y LOBKOWITZ, Juan. 1678. *Arquitectura civil recta y oblicua ...* Vigevano: Imprenta obispal.

CLAIRAUT, Alexis Claude. 1731. *La Recherche sur les courbes à double courbure.* Noyon, Didot & Quillaut.

CURABELLE, Jacques. 1644. *Examen des oeuvres du Sieur Desargues.* Paris: M. & I. Henault - F. L'Anglois dit Chartres.

DE L'ORME, Philibert. 1561. *Nouvelles inventions pour bien bastir a petits frais.* Paris: Federic Morel.

DERAND, François, S. I. 1643. *L'Architecture des voûtes ou l'art des traits et coupe des voûtes ...* Paris: Sébastien Cramoisy.

DESARGUES, Girard. 1640. *Brouillon project d'exemple d'une manière universelle du S.G.D.L. touchant la practique du trait a preuues pour la coupe des pierres en l'architecture* ... Paris: Melchoir Tavernier.

———. 1648. Reconnoissance de Monsieur Desargues. In: Abraham BOSSE, *Maniere universelle de Mr. Desargues, pour pratiquer la perspective par petit-pied, comme le geometral. Ensemble les places et proportions des fortes & foibles touches, teintes ou couleurs*. Paris: Imprimerie de Pierre Des-Hayes.

DÜRER, Albrecht. 1525. *Underweysung der messung mit dem zirkel und richtscheyt in Linien ebnen unnd gantzen corporen durch Albrecht Dürer zu samen getzogen zu nutz aller kunstliebhabenden mit zuu gehörigen figuren in truck gebracht im jar MDXXV.* Nuremberg: s.n. Electronic facsimile c. 2008. Firenze:Instituto e Museo di Storia della Scienza. Available from http://bibdig.museogalileo.it/rd/bd.

EISENMANN, Peter, Michael GRAVES, Charles GWATHMEY, John HEJDUK and Richard MEIER. 1975. *Five architects: Eisenman, Graves, Gwathmey, Hejduk, Meier.* New York: Oxford University Press.

ETLIN, Richard. 2009. Génesis y estructura de las bóvedas de Arles. Pp. 425-434 in *Actas del Sexto Congreso Nacional de Historia de la Construcción.* Madrid: Instituto Juan de Herrera.

EULER, Leonhard. 1772. De solidis quorum superficiem in planum explicare licet (E419). *Novi Commentarii academiae scientiarum Petropolitanae* 16: 3-34. Rpt. *Leonhardi Euleri Opera Omnia,* Series I, vol. 28, pp. 161-186.

FARISH, William. 1822. On Isometrical Perspective. *Transactions of the Cambridge Philosophical Society* 1: 1-19.

FREZIER, Amédée-François. 1737-1739. *La théorie et la pratique de la coupe des pierres et des bois ... ou traité de stéréotomie a l'usage de l'architecture.* Strasbourg-Paris: Jean Daniel Doulsseker-L. H. Guerin.

GERBINO, Anthony. 2002. François Blondel (1618-1686): Architecture, Erudition and Early Modern Science. Ph.D. thesis. New York, Columbia University.

———. 2005. François Blondel and the Résolution des quatre principaux problèmes d'architecture (1673). *Journal of the Society of Architectural Historians* 64, 4: 498-521.

GIL DE HONTAÑÓN, Rodrigo. 1540 c. "Manuscrito", included in Simón GARCÍA, "Compendio de Arquitectura y simetría de los templos" (1681), MS 8884, Biblioteca Nacional, Madrid.

GÓMEZ MARTÍNEZ, Javier. 1998. *El gótico español de la Edad Moderna. Bóvedas de Crucería.* Valladolid: Universidad.

GÓMEZ-FERRER LOZANO, Mercedes and Arturo ZARAGOZÁ CATALÁN. 2008. Lenguajes, fábricas y oficios en la arquitectura valenciana del tránsito entre la Edad Media y la Edad Moderna (1450-1550). *Artigrama* 23: 149-184.

GUARINI, Guarino. 1737. *Architettura Civile del padre D. Guarino Guarini chierico regolare opera postuma dedicata a sua sacra reale maesta.* Torino: Gianfrancesco Mariesse.

HACHETTE, Jean Nicolas Pierre. 1822. *Traité de géométrie descriptive: comprenant les applications de cette géométrie aux ombres, à la perspective et à la stéréotomie.* Paris: Guillaume et Corby.

HASELBERGER, Lothar. 1983. Die Bauzeichnungen des Apollontempels von Dydima. *Architectura* 13, 1: 13-26.

———. 1994. The Hadrianic Pantheon – a Working Drawing Discovered. *American Journal of Archaeology* 98, 2: 327.

HUERTA FERNÁNDEZ, Santiago. 2007. Oval Domes: History, Geometry and Mechanics. *Nexus Network Journal* 9, 2: 211-248.

JOUSSE, Mathurin. 1642. *Le secret d'architecture découvrant fidélement les traits géométriques, couppes et dérobements nécessaires dans les bastimens.* La Flèche: Georges Griveau.

LA GOURNERIE, Jules-Antoine-René Maillard de. 1851. Considerations géométriques sur les arches biaises. *Annales des Ponts et Chaussees*: 82-115.

———. 1853. Note sur les arches biaises a l'occasion du memoire de M. Graeff. *Annales des Ponts et Chaussees*: 281-288.

———. 1855. *Discours sur l'art du trait et la Géometrie Descriptive ... prononcé au Conservatoire Impérial des Arts et Metiers, le 14 novembre 1854, a l'ouverture du cours de geometrie descriptive* ... Paris: Mallet-Bachelier.

————. 1860. *Traité de géométrie descriptive* ... Paris: Mallet-Bachelier.

————. 1872. Mémoire sur l'appareil de l'arche biaise, suivi d'une analyse des principaux ouvrages publiés sur cette question et d'une reponse a des critiques sur l'enseignement de la stéréotomie a l'Ecole Polytechnique. *Annales du Conservatoire des arts-et-métiers* **9**, 1ère série: 332-406.

————. 1874. Mémoire sur l'enseignement des Arts graphiques. *Journal de Mathématiques Pures et Apliquées* **19**, 2ème serie: 113-156.

LA RUE, Jean-Baptiste de. 1728. *Traité de la coupe des pierres où par méthode facile et abrégée l'on peut aisément se perfectionner en cette science.* Paris: Imprimerie Royale.

LAUSSEDAT, Aimé. 1897. De la Gournerie (1814-1883). Pp. 135-140 in *Ecole polytechnique. Livre du Centenaire.* Paris: Gauthier-Villars et fils.

LEROY, Charles-François-Antoine. 1834. *Traité de géométrie descriptive, avec une collection d'épures composée de 60 pl.* ... Paris: Carilian-Goeury.

LÓPEZ DE ARENAS, Diego. 1633. *Breve compendio de la carpintería de lo blanco, y tratado de alarifes.* Sevilla: Luis Estupiñán.

LÓPEZ MOZO, Ana. 2009. Bóvedas de piedra del Monasterio de El Escorial. Ph.D. thesis, Universidad Politécnica de Madrid.

LORIA, Gino. 1921. *Storia della Geometria Descrittiva, dalle origini sino ai giorni nostri.* Milan: Ulrico Hoepli.

————. 1567. *Le premier tome de l'Architecture.* Paris: Federic Morel.

MARTÍNEZ DE ARANDA, Ginés. 1600 c. *Cerramientos y trazas de montea* (ms). Madrid, Biblioteca del Servicio Histórico del Ejército. (Facsimile ed. Madrid: CEHOPU, 1986).

MILLIET DE CHALLES, Claude-François, S. I. 1674. *Cursus seu mundus mathematicus.* Lugduni: Officina Anissoniana.

MONGE, Gaspard. 1769. Sur les développées des courbes à double courbure et leurs inflexions. *Journal encyclopédique*: 284-287.

————. 1795 [An III]. Stereotomie. *Journal de l'Ecole Polytechnique* **1**: 1-14.

————. 1796 [An IV]. Des lignes de courbures de la surface de l'Ellipsoïde. *Journal de l'Ecole Polytechnique* **2**: 145-165.

————. 1798 [An VII]. *Géométrie descriptive, leçons données aux Écoles normales, l'an 3 de la République* ... . Paris: Baudouin.

MONGE, Gaspard and Barnabé BRISSON. 1820. *Géométrie descriptive, par G. Monge. 4e édition augmentée d'une théorie des ombres et de la perspective, extraite des papiers de l'auteur, par M. Brisson.* Paris: Vve Courcier.

MÜLLER, Werner. 1990. *Grundlagen gotischer Bautechnik. Ars sine scientia nihil est.* Munich: Deutscher Kunstverlag.

OLIVIER, Théodore. 1843-1844. *Cours de géométrie descriptive.* Paris: Carilian-Goeury et V. Dalmont.

————. 1852. *Mémoires de géométrie descriptive.* Paris: Carilian-Goeury et V. Dalmont.

PALACIOS GONZALO, José Carlos. 1987. La estereotomía de la esfera. *Arquitectura* **267**: 54-65.

————. 1990. *Trazas y cortes de cantería en el Renacimiento Español.* Madrid: Instituto de Conservación y Restauración de Bienes Culturales. Ministerio de Cultura.

PELLEGRINO, Alba Ceccarelli. 1996. *Le "Bon architecte" de Philibert de l'Orme hypotextes et anticipations.* Fasano-Paris: Schena-Nizet.

PEROUSE DE MONTCLOS, Jean-Marie. 1982. *L'Architecture a la française.* Paris: Picard.

PIERO DELLA FRANCESCA. 1480 c. *De prospectiva pingendi.* (Facsimile ed. with commentary, Massimo Mussini and Luigi Grasselli, eds. Arezzo: Aboca Museum Editions, 2008).

PONCELET, Jean-Victor. 1822. *Traité des propriétés projectives des figures ouvrage utile à ceux qui s'occupent des applications de la géométrie descriptive et d'opérations géométriques sur le terrain.* Paris: Gauthier-Villars.

PORTOR Y CASTRO, Juan de. 1708. *Cuaderno de arquitectura.* MS 9114, Biblioteca Nacional, Madrid.

POTIÉ, Philippe. 1996. *Philibert de l'Orme. Figures de la pensée constructive.* Marseille: Éd. Parenthèses.

RABASA DÍAZ, Enrique. 1996a. Arcos esviados y puentes oblicuos. El pretexto de la estereotomía en el siglo XIX. *Obra Pública* **38**: 18-29.

———. 1996b. Técnicas góticas y renacentistas en el trazado y la talla de las bóvedas de crucería españolas del siglo XVI. Pp. 423-434 in *Actas del Primer Congreso Nacional de Historia de la Construcción*. Madrid: Instituto Juan de Herrera.

———. 1999. Auguste Choisy: vida y obra. Pp. xxiii-xxiv in *El arte de construir en Roma*. Madrid: Instituto Juan de Herrera.

———. 2000. *Forma y construcción en piedra. De la cantería medieval a la estereotomía del siglo XIX*. Madrid: Akal.

———. 2007. 'Plomo' y 'nivel': Hábitos de pensamiento espacial en la construcción gótica. in *La piedra postrera. Simposio Internacional Sobre la Catedral de Sevilla en el Contexto del Gótico Final. V Centenario de la Conclusión de la Catedral de Sevilla*. Sevilla: Cabildo de la Santa, Metropolitana y Patriarcal Iglesia Catedral de Sevilla-Escuela de Estudios Árabes. CSIC.

RECHT, Roland. 1989. Les 'traités pratiques' d'architecture gothique. Pp. 279-285 in *Les bâttisseurs des cathédrales gothiques*. Strasbourg: Editions Les Musées de la Ville de Strasbourg.

———. 1995. *Le dessin d'architecture. Origine et fonctions*. Paris: Adam Biro.

ROJAS, Cristóbal de. 1598. *Teórica y práctica de fortificación, conforme a las medidas de estos tiempos* ... Madrid: Luis Sánchez.

RORICZER, Mathes. 1486. *Büchlein von der fialen Gerechtigkeit*. Regensburg.

———. 1490 c. *Geometria Deutsch* (includes four sheets on gablets that are considered by Lon Shelby to be an independent booklet).

RUIZ DE LA ROSA, José Antonio and Juan Clemente RODRÍGUEZ ESTÉVEZ. 2002. 'Capilla redonda en vuelta redonda' (sic): Aplicación de una propuesta teórica renacentista para la catedral de Sevilla. Pp. 509-516 in *IX Congreso Internacional Expresión Gráfica Arquitectónica. Revisión: Enfoques en docencia e investigación*. La Coruña: Universidad de A Coruña.

RUIZ EL JOVEN, Hernán. 1550 c. *Libro de Arquitectura* (ms). Madrid, Biblioteca de la Escuela de Arquitectura de la Universidad Politécnica de Madrid.

HUGH OF SAINT VICTOR. 1991. *Practical Geometry (Practica geometriae)*. Frederick A. Homann, ed. and trans. Milwaukee: Marquette University Press.

SAKAROVITCH, Joël. 1992. La coupe des pierres et la géometrie descriptive. Pp. 530-540 in *L'Ecole Normale de l'an III Leçons de Mathématiques, Laplace-Lagrange-Monge*. Paris: Dunod.

———. 1994a. La géométrie descriptive, une reine déchue. Pp. 77-93 in *La formation polytechnicienne, deux siècles d'histoire*. Paris: Dunod.

———. Le fascicule de stéréotomie; entre savoir et metiers, la fonction de l'architecte. Pp. 347-362 in *Desargues en son temps*. Paris: Blanchard.

———. 1995. The Teaching of Stereotomy in Engineering schools in France in the XVIIIth and XIX centuries: an Application of Geometry, an 'Applied Geometry', or a Construction Technique?. Pp. 204-218 in *Entre mécanique et architecture/Between Mechanics and Architecture*. Patricia Radelet-De Grave and Edoardo Benvenuto, eds. Basel-Boston-Berlin: Birkhäuser.

———. 1997. *Epures d'architecture*. Basel-Boston-Berlin: Birkhäuser.

———. 2009a. Gaspard Monge Founder of 'Constructive Geometry'. Pp. 1293-1300 in *Proceedings of the Third International Congress on Construction History*. Cottbus: Brandenburg Technical University.

———. 2009b. From One Curve to Another or the Problem of Changing Coordinates in Stereotomic Layouts. Pp. 297-319 in *Creating Shapes in Civil and Naval Architecture. A Cross-Disciplinary Comparison,* Horst Nowacki and Wolfgang Lefèvre, eds. Leiden-Boston: Brill.

SAN NICOLÁS, Fray Laurencio de. 1639. *Arte y uso de Arquitectura*. s. l.: Imprenta de Juan Sanchez.

SANABRIA, Sergio Luis. 1984. The evolution and late transformations of the Gothic mensuration system. Ph.D. Thesis, Princeton University.

SERLIO, Sebastiano. 1545. *Il primo libro d'architettura* ... Paris: Iehan Barbé.

SHELBY, Lon R. 1969. Setting Out the Keystones of Pointed Arches: A Note on Medieval 'Baugeometrie'. *Technology and Culture* **10**, 4:537-548.

———. 1972. The geometrical knowledge of medieval master masons. *Speculum* **47**, 3: 395-421.

SPENCER, H. 1839. Skew arches. *The Civil Engineer and Architect's Journal* 2: 164-165.

SWANSON, Randy S. 2003. Late XVIIth century practice of stereotomy prior to the establisment of Engineering Schools in France. Pp. 1875-1885 in *Proceedings of the First International Congress on Construction History*. Madrid: Instituto Juan de Herrera.

TAÍN GUZMÁN, Miguel. 1998. *Domingo de Andrade, Maestro de Obras de la Catedral de Santiago (1639-1712)*. Sada-A Coruña: Ediciós do Castro.

TAMBORÉRO, Luc and Joël SAKAROVITCH. 2003. The vault at Arles City Hall: A carpentry outline for a stone vault?. Pp. 1899-1907 in *Proceedings of the First International Congress on Construction History*. Madrid: Instituto Juan de Herrera.

TATON, René. 1954. *L'Histoire de la géométrie descriptive*. Paris: Université de Paris.

TOMLOW, Jos. 2009. On Late-Gothic Vault Geometry. Pp. 193-219 in *Creating Shapes in Civil and Naval Architecture. A Cross-Disciplinary Comparison*, Horst Nowacki and Wolfgang Lefèvre, eds. Leiden-Boston: Brill.

TOSCA, P. Thomas Vicente. 1707-1715. *Compendio mathemático, en que se contienen todas las materias más principales de las Ciencias, que tratan de la cantidad....* Valencia: Antonio Bordazar - Vicente Cabrera.

VANDELVIRA, Alonso de. 1585 c. *Libro de trazas de cortes de piedras*. Ms. R31, Biblioteca de la Escuela Técnica Superior de Arquitectura de la Universidad Politécnica de Madrid.

VILLARD DE HONNECOURT. 1225 c. *Carnet*. MS fr 19093, Bibliothèque Nationale de France, Paris.

———. 2009. *The Portfolio of Villard de Honnecourt, A New Critical Edition and Color Facsimile*. Carl F. Barnes, ed. Farnham: Ashgate.

WILKINSON ZERNER, Catherine. 1977. The New Professionalism in the Renaissance. Pp. 124-153 in *The Architect. Chapters in the History of the Profession*. New York: Oxford University Press.

## About the author

José Calvo-López is an architect. His Ph.D. dissertation, about the stonecutting manuscript of Ginés Martínez de Aranda, was awarded the Extraordinary Doctoral Prize of the Polytechnic University of Madrid in 2001. He is Professor of Graphical Geometry at the School of Architecture and Building Engineering of the Polytechnic University of Cartagena. He has also lectured on stonecutting, stereotomy, the history of spatial representation and photography at the Polytechnic Universities of Valencia and Madrid and at San Pablo-CEU University. His research, focused on stereotomy and other issues concerned with spatial representation, is published regularly in refereed journals, international conferences, and such books as *Cantería renacentista en la catedral de Murcia* (Murcia: Colegio Oficial de Arquitectos de Murcia, 2005).

José Carlos Palacios Gonzalo[*]
*Corresponding author

Escuela Técnica Superior
de Arquitectura de Madrid
Universidad Politécnica de Madrid
Avda. Juan de Herrera 40
Madrid 28040 SPAIN
josecarlos.palacios@upm.es

Rafael Martín Talaverano

Escuela Técnica Superior
de Arquitectura de Madrid
Universidad Politécnica de Madrid
Avda. Juan de Herrera 40
Madrid 28040 SPAIN
guinzip@gmail.com

Keywords: Cross ribbed vault, ribs, geometry,
semicircular arch, pointed arch, oval arch, ridge
line, tierceron, formeret, transverse arch,
voussoir, boss stone, stonecutting, stereotomy,
Juan de Álava, Rodrigo Gil

Research

# Geometric Tools in Juan De Álava's Stonecutting Workshop

**Abstract.** Stonemasonry of the Gothic vault in its totality is based upon geometry of the line, whereas classic stereotomy relies on the comprehensive knowledge of the surface and the highly sophisticated sides of the voussoirs necessary for its vaults. It is obvious that this leap in the art of construction was paralleled and accompanied by an extension of the horizons of geometry. In Spain, it was made possible thanks to the centuries-old tradition of stone building begun in the most remote medieval times and to the presence of outstanding architects or stonemasons such as Juan de Álava, whose professional work surpassed the established limits and provided the art of building with new instruments.

## 1 Juan de Álava

Medieval stonecutting workshops were places of an extraordinary interest; stonecutters not only tackled the cutting of the various architectural pieces intended for the buildings under construction but also, and more importantly, gave rise to modern geometry. The Gothic vault was originally a rather simple structure, solved with a ribbed crossing, that is, two diagonal arches which intersect in the centre of the vault; nevertheless, this vault, the undisputed main feature of classic French Gothic, took a turn towards complexity. The number of arches multiplied, and so did their intersections on the surface of the vault; the vault began to assume complex forms as a consequence of the web of ribs on which the vault was based. Stonemasons, facing the geometric issues resulting from modern ribbed vaults, had to look for geometric solutions which would allow them to solve the increasingly complex shapes of Gothic vaults: a geometry that was capable of inferring the vault volume from its horizontal projection.

In the first half of the sixteenth century, when we first hear of Juan de Álava (1480? - 1537), Spanish Gothic is at its summit. This architect is undoubtedly one of the outstanding personalities in Spanish Gothic, a style that, far from disappearing in the Renaissance, attained its greatest moment in Spain. Álava's architectural production is considerable and, in particular, achieved a remarkable repertoire of vault like forms which always show his very personal design. When designing his vaults, Juan de Álava avoids the classical Castilian quatrefoil, creating instead a web of ribs that extends along the aisles. This type of vault, completely original in the Iberian Peninsula, is designed only in Álava's workshop.

To build these complex ribbed vaults, this renowned architect resorted to extraordinarily interesting geometric solutions with which he confers shape on a set of architectural shells, one of the most surprising aspects of his work. This article intends to show the advances in geometry achieved by Juan de Álava's construction method. At the beginning of the sixteenth century, the dihedral system of projection was about to make its appearance, and the majority of the tools used for its calculation were already known.

## 2 Gothic architecture: evolution

The classic Gothic vault is a consequence of the groined vault. Let us recall that the Roman groined vault results from the intersection of two semi-cylinders and, as everybody knows, their crossing produces two elliptical arches. When the vault is made with concrete or rubble masonry, costly wooden centerings need to be built so as to pour on them the material with which the vault is constructed; the elliptical lines of the cross are the natural result of the line left by the two centerings when mortar is poured on them. In the Romanesque style, as in the case of the crypt under the Portico of Glory in Santiago de Compostela, they indulged in highlighting these lines with stone ribs so as to reinforce them, but these reinforcements are absolutely unnecessary in vaults made with rubble masonry. When building the vault with bonded stone, the ashlar beds of the two barrels produce a groined joint along this intersection; in order to solve this joint, there are two options: either to bond the stones which form the groin lines in two directions, entailing extremely difficult stonecutting, or hide this joint with a diagonal rib.

The Gothic vault is the outcome of a variety of achievements occurring simultaneously, of which one takes place when Gothic architects realize that resorting to diagonal ribs, combined with the use of transverse arches as resistant elements, not as mere decoration, can make the very expensive centerings required by Roman and Romanesque vaults unnecessary. The vault can be closed by covering the spaces between the ribs, that is to say the panel, with pieces of smaller vaults without any kind of centering, since the cells start taking shape through small arches that are completely self-bearing and placed between two adjoining ribs. Later in time, with the appearance of new ribs, the vault was broken up into smaller compartments and the panel could be made almost flat between the ribs, making its construction easier and the web considerably thinner. The thick shells of Early Gothic became a thing of the past. In short, the presence of ribs provided an easy solution to the difficult diagonal joint, allowed savings in the panel centerings that can be solved with simple incidental supports and, lastly, represented a powerful aesthetic solution.

A new problem arose: the difficulty encountered when building an elliptical arch in stone since all its voussoirs are different. Let us add that when the vault has a rectangular plan, both barrels have a different span and, in order to keep the vault's ridge line horizontal, the smaller barrel has to be stilted with the unfortunate consequence that the joint of the barrels – the ellipse – no longer occurs in a vertical plane. This circumstance made it necessary to build an elliptical arch whose outline occupies three dimensions. This geometrical problem, difficult from the point of view of construction, was successfully solved by replacing the ellipse with two semicircular arches placed in the vertical plane. However, this choice which, in a certain way, entails a distortion of geometry, has its consequences: first, the vault's crown is higher, so in order to maintain its ridge line horizontal, the semi-cylindrical barrels have to be pointed; second, the vault shells which were cylindrical in the traditional groined vault are no longer so, because, between the curvature of the diagonal arches and the four pointed wall ribs, are formed ruled surfaces shaped like a plow blade, which have to be built with the bonded stones that make up the cell [Frankl 1962: 80; Bechmann 1981: 159; Choisy 1899: 271, 274].

It is known that when the span of the aisles began to reach considerable dimensions – near 15 m. as in most French cathedrals from the thirteenth century on – the diagonal arches are pointed, as Violet-le-Duc explained, so as to reduce strong thrust caused by semi-circular diagonal arches. However, formeret and transverse arches generally keep their original heights so as not to increase even further the height of the perimeter of the aisles, resulting in a vault with slightly inclined ridge lines.

Indeed, the classical Gothic vault was the result of a series of intelligent decisions which led to vaults that were easy to build, economical and satisfactory from an aesthetic perspective. Later on, appeared the tierceron, whose contructive rationale, although justified by Viollet-le-Duc as an elbow-shaped reinforcement of the lintel rib, in fact most likely responds to the need to fragment the cells in order to diminish the length of the beds, thus facilitating its construction and make it less expensive.

Needless to say, from an aesthetic point of view the five-boss vault was a great success compared to the groined vault, and with it the surface of the Gothic vault began to be fragmented into individual compartments, or cells, separated by a grid of ribs.

## 3 From the semicircular arch to the pointed arch

Let us talk about some advantages of building with diagonal arches. When trying to cover a certain span with a semi-circular arch, there is just one solution: the voussoirs forming that arch will be determined by the radius of curvature of the circumference that describes it. If another arch with a different span is required at the same location, the voussoirs for that second arch would have to have another radius of curvature, and, therefore, they would be different from those of the previous arch; this would be the case with all the arches of distinct span. Consequently when building with semicircular arches, the voussoirs' arrangement, storage and assembly becomes more and more complicated, since the pieces are not interchangeable among themselves; each one requires its adequate placing.

Nevertheless, if, to attain the same span, the arch adopted was pointed instead of semi-circular, the solution is indeterminate, since there are infinite kinds of pointed arches that could cover a given span. It is therefore compulsory to define another parameter to determine the solution, and in medieval times that was the height of the keystone, which made it possible to decide the radius and the centre of curvature, presuming that the arch springs vertically. The degree of indetermination inherent in the geometry of the pointed arch is, at the same time, one of its most powerful advantages. If the plan is to build another arch to cover a different span, it is possible to draw another figure with the same radius of curvature, and, in this case, the variation occurs in the height of the keystone. It is, therefore, possible to draw several arches to cover different spans with the same radius of curvature; in other words, the same voussoirs can be used for different arches. Fig. 1 shows a drawing by Viollet-le-Duc in which this principle is summarized: with the same curvature of a semi-circular arch several pointed arches can be drawn. Applying this principle to a star-shape ribbed vault in which the diagonal arch is a semi-circular arch, the other arches, the tierceron and the formeret could be built with identical curvatures.

Therefore, the use of the pointed arch entails a remarkable advantage for the organization of the works and the transmission of orders among the different members of the workshop constructing the building.

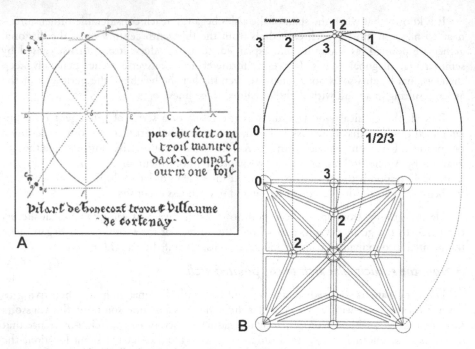

Fig. 1. Drawing by Viollet-le-Duc from a sketch of Villard de Honnecourt, showing how all ribs of a ribbed vault can be drawn from the semicircular arch of the diagonal rib [Viollet-le-Duc 1878: 130])

## 4 Complexity and standardization

Gothic was an architectural style which succeeded in generating a very articulated aesthetic code, but in addition to its stylistic singularities, it is interesting to notice how the composition rules which made up this style are justified by the existence of a vector which, in a continuous way , encouraged the style to attain higher and higher degrees of complexity. As far as vault building is concerned, everybody knows that Gothic style represents the success of the cross-ribbed vault, undoubtedly one of the most revolutionary inventions in construction history. It is the result of a long process which goes from the simple crossing of ribs in vaults at the beginning of the thirteenth century to the spectacularly complex late Gothic vaults of the sixteenth century.

The principle of complexity as a vector of Gothic evolution was an idea developed in the middle of the twentieth century by Paul Frankl [1962]. This present article aims at elaborating on this issue by adding that the large complexity attained by European Gothic was due to the tools generated in the workshops of medieval stonemasons: geometry and standardization. The aim was to attain the largest complexity with the widest standardization, so as to succeed in building the most complicated ribbed vault with just one radius. The advantages of this technique are obvious: all the voussoirs with which the ribs are built are identical, and so is the wooden centering needed for their construction.

To build a vault's shell by collecting ribs necessitated a strict control of several points in the space which can not be carried out without a largely developed geometry. Each arch has a springing point and must reach a height which is determined by the surface to be built, both points are indispensable to carring out the layout of each arch.

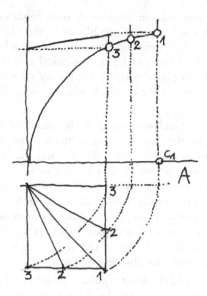

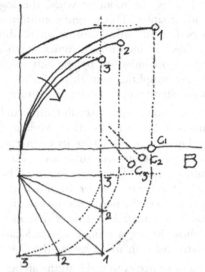

Fig. 2A. All the arches are equal because they coincide with the curvature of the diagonal

Fig. 2B. The arches are equal because the diagonal arch (semi-circular) bends forward until it reaches the height of the secondary keystones previously determined

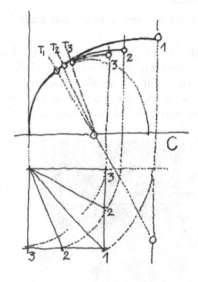

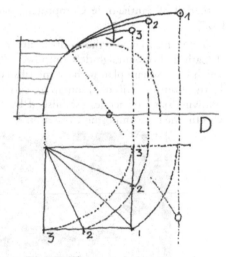

Fig. 2C. When using oval arches, the upper portion of the oval can slide down the lower part. The point of tangency between the two curvatures of the oval moves, but all the arches can be built with the same curvatures

Fig. 2D. The previous drawing can be simplified even more. The lower part of the oval arch does not really exist, it is the solid springing of the vault. The upper part of the oval starts always at the same height, where the springing finishes. From this point, as in case B, the arch formed by the upper part of the diagonal oval tilts to reach the heights of the boss

Fig. 2. Summary sketches containing four different techniques detected in the Spanish Gothic vaults in order to standardize the curvatures of the arches in a vault

This is the moment when the logic of construction imposes a new condition: standardization. This second condition requires the increasing complexity in the ribs to be performed with a minimum of different radii, even with just one radius, when possible. The increasing complexity inherent in the Gothic style, as paradoxical as it may seem, is accompanied by the aim to attain the largest degree of complexity with the greatest simplification; this principle which cannot be applied without achieving standardization of the arches' curvatures.

The evolution of Spanish Gothic architecture in its latest period, in the middle of the transition between the Late Middle Ages and the modern era, is marked by the development of ribbed vaults of increasing complexity in which the ribs and the boss which form them are multiplied incessantly. The most usual model is the stellar groin vault, formed of five bosses (a central boss and four secondary). This composition is the basis of a great number of much more complex compositions in which both the keystones and the tiercerons increase in number to attain compositions that are more and more spectacular. As a general rule, most of these vaults use compass-drawn arches, that is, those having a unique centre located on the impost line of the vault, but vaults constructed with arches having three centers, i.e., ovals, are also very frequent.

The strategies to make all the arches used to build the vault equal are summarized up in the four sketches shown in fig. 2. Let's see now in greater detail the application of each of these cases in the vaults of Juan de Álava.

## 5 The vaults of "rampante plano" (flat rampant)

### Cathedral of Santiago de Compostela, vaults of the cloisters. North nave: built 1521-1527

Let us now analyze the vaults built by Juan de Álava in the north aisle of the cloister of Cathedral of Santiago de Compostela [Castro 2002; Gómez Martínez 1998] (fig. 3)., which has a square plan with all the ribs equal, that is, with identical curvature. After his death, Rodrigo Gil de Hontañón carried on with the construction of the cloister following same principles established by Juan de Álava, as did the successive architects who finished this imposing cloister on the south aisle.

Fig. 3. Cathedral of Santiago de Compostela, cloister, north aisle, Juan de Álava, 1521-27

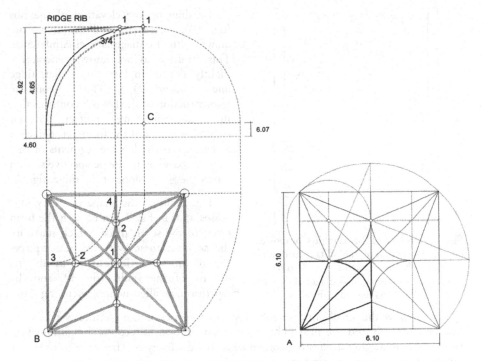

Fig. 4. Cathedral of Santiago de Compostela. Drawing of a cloister vault.
A) plan; B) elevation of the curvatures of its ribs

The plan is a square with a side measuring 6.10 m; over it there is a simple ribbed design which, as we will see, turns out to be spectacular (fig. 4A). The tiercerons are located in a intermediate position between the diagonal and the formeret arches; thus, once the positions of the bosses of the tiercerons are established, vertical and horizontal lines make it possible to determine, on the diagonal lines, the centres of the curves of the subsidiary ribs. Once drawn, they describe a square with concave sides in the centre of the vault. In addition to these ribs, four lierne ribs highlight the ridge line.

The systematic standarization of the sections of the ribs is a remarkable feature in Juan de Álava's vaults [García 1991]. The ribs in this vault all have the same section (around 20 cm), except for the transverse arch, which is slightly thicker (about 25 cm).

Having ribs of equal section help unify the individual bays into a larger whole, the centralized design of each bay becomes less important and the fragmentation into parts is diminished; as a consequence, the whole aisle gives an impression of unity. This visual effect can be stressed by a design of subsidiary ribs linking the vaults. In the cloister of the cathedral of Compostela this is the function carried out by the central lozenge: its curvature, after a straight part of the lierne, goes on in the next part, and so on.

First of all, just by looking at the springing of the vault one can notice a rather marked 60 cm. stilt; second, all ribs are aligned in their intrados; because they are all equal, the aesthetic effect is that of a fluted column. The intelligent combination of both resources creates that impression of balanced classicism derived from this cloister.

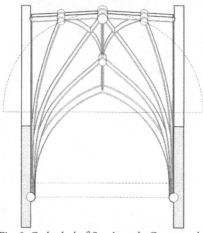

Fig. 5. Cathedral of Santiago de Compostela. Prospective section of one of its cloister vaults in which it is allowed to see its stilted springing, its semicircular diagonal rib and, also, its flat ridge lines

By drawing the elevation of the ribs (fig. 4B), we see that all the arches are drawn with the diagonal's circumference. This strategy, as we know, generates a slightly sloped, in fact almost flat, ridge line (*rampant*). The geometric reconstruction of the lay-out confirms that the diagonals are perfect circular circumferences and the formeret, tierceron and transverse arches are segments of that large diagonal arch; the perspective section shows the gentle slope of the ridge (fig. 5).

Fig. 6 shows the shape of this vault's bosses and, in fig. 7 can be seen the form given to the seven pieces which make up the *tas-de-charges*. With this set of pieces (fig. 8) can be carried out the assembling of one of the vaults (fig. 9), and, by repetition, an alignment of them (fig. 10).

We see how the small concave square works to generate the continuity of the ribs between one vault and the next. The flat rampant in both directions, longitudinal and transversal, has another interesting consequence: the discovery of a fan vault in its simpler configuration: the square trumpet-bell shape.

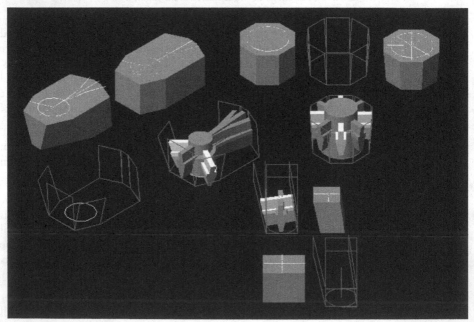

Fig. 6. Cloister vaults in the Cathedral of Santiago de Compostela. In order to cut the boss stones it is necessary to draw them in plan. The horizontal projection of the boss stone is drawn on a flat face of the stone prism; after that, the main shape of the boss stone is obtained by carving away the stone outside the plan drawing

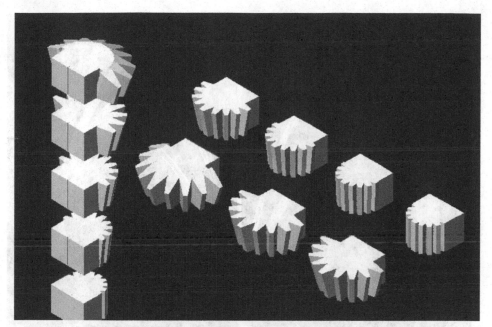

Fig. 7. Cathedral of Santiago de Compostela, cloister.
The *tas-de-charges* is composed of seven blocks of dressed stone

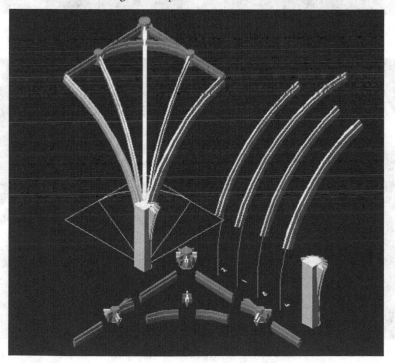

Fig. 8. Cathedral of Santiago de Compostela, a cloister vault. 3D reconstruction of the different
pieces that compose a quarter of the vault

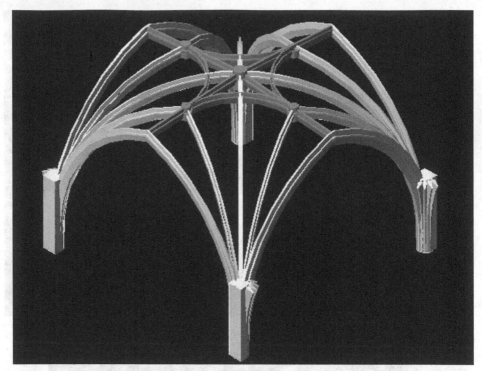

Fig. 9. Cathedral of Santiago de Compostela. Modeling of one of the cloister vaults, in which the specifics features of this vault are clearly shown

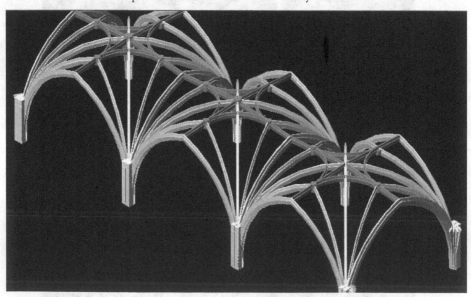

Fig. 10. Cathedral of Santiago de Compostela, cloister. Alignment of the vault. The flat ridge line results in square trumpet-bell shapes, and the subsidiary ribs link the vaults to each other

Fig. 11. Cathedral of Santiago de Compostela, cloister. The corner of the cloister, where the subsidiary ribs enhance the continuity of the vaults

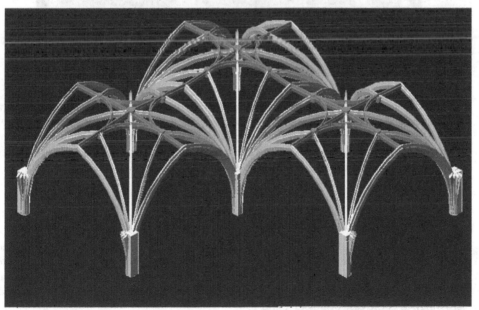

Fig. 12. Cathedral of Santiago de Compostela, cloister. 3D model of a vault turning a corner. All elements with which the vault is composed collaborate to enhance the expression of continutity

Juan de Álava was aware of the formal expressivity of this element, as shown in the corner of the cloister where the vault turns the corner in a most beautiful and elegant way (figs. 11, 12). To our surprise, the architect renounced the opportunity to reinforce this conic effect with horizontal courses that would have created circular stone beds around a vertical axis. Instead, he maintained the Spanish construction tradition of panel bonding and placed the stones *á la française*, that is, as in a groined vault. The cells are made with large stones which span the intervals between the ribs in single pieces, thus requiring therefore no centering or ancillary means of support during construction.

## 6 Pointed barrel vaults

### San Esteban's Monastery, Salamanca. Nave vault. Started in 1526-1533

The vault covering the nave in San Esteban's convent is one of the most spectacular vaults built in Spain. Juan de Álava's masterpiece is an example of his specific way of conceiving the Gothic vault [Castro 2002; Chueca 2001; Navascués 1984] (fig. 13).

Fig. 13. Convent of San Esteban, Salamanca, vaulting of the nave by Juan de Álava, 1526-33

This vault of this convent church, with a span of 14.50 m, is as wide as the nave of a large cathedral, even exceeding that of Salamanca (13 m). Let us recall that the cathedral of Plasencia, at 17.30 m, has the third largest span in Spanish Gothic, following those of Gerona (23 m) and Palma (20 m). We can therefore attribute to Juan de Álava the merit of having achieved some of the largest spans in Spanish Gothic. After Álava's death, the construction of both Plasencia and San Esteban was carried on by Rodrigo Gil de Hontañón.

The nave of San Esteban is divided into segments of roughly 7 m, so each segment is a rectangle *duplo*, that is, twice as long as it is wide. The complex design of its rib work is perfectly determined in an orthogonal 8 x 4 grid (fig. 14), making it possible to fix the position of the double bosses of the tiercerons along the major axis and of the simple tiercerons along the minor axis. This grid also determines the dimension of the central circle. Drawing from these points, by alignments, are obtained the other bosses and crossings, as is the case of bosses 4 and 2.

When calculating the curvatures of the arches (fig. 14B), the skill with which this vault is designed becomes evident. We soon realize that its diagonal arch is exactly a circular circumference with its horizontal plane at the height of the capitals.

Later, when drawing the *rampants*, we were able to confirm that, towards the major axis, the ridge line is slightly curved and presents a steep slope to the formeret arch of almost 2 m high; in contrast, towards the minor axis, the ridge line is almost flat. As we know, this combination of horizontal *rampants* along the nave axis and a steep slope in the transverse direction, is what creates the tunnel effect in this type of vaults.

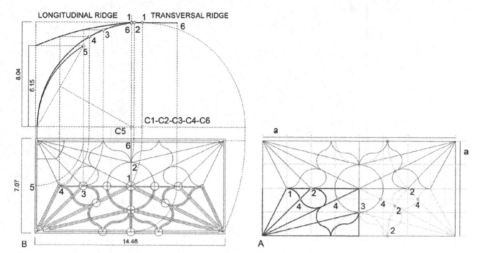

Fig. 14. Convent of San Esteban, Salamanca. Drawing of a cloister vault: A) plan; B) arches curvatures elevation. Note that the curvatures of its arches coincidence perfectly with the semicircular diagonal arch; only the curvature of the formeret arch seems to be different

When calculating the elevation of the other arches, to our surprise, we realized that tiercerons 4 and 3, as well as tierceron 2 and the transverse arch, coincide with the semi-circular circumference of the diagonal (fig. 14B); the only different arch is formeret 5. As frequently in Álava's work, the shape of the transverse pointed arch comes very close to being a semi-circular circumference; its centre is situated next to the arch's vertical axis, at approximately an eighth of half of its span. Therefore, in spite of this vault's apparent complexity, all its main ribs have identical curvatures. Here again we find an insistence on systematizing and standardizing the form of the arches.

The three-dimensional drawings of a segment in particular (fig. 16) and of the nave as a whole (fig. 17) allows us to visualize the huge spatial web made up by the bosses and ribs; these three-dimensional grids are the core of the vaulting works of late Gothic. Upon them are easily laid out the thin panels which, as regards the work of Álava, are always bonded *á la française*.

The *combados*, or subsidiary ribs, play a very important role. It is an extremely original configuration which has an influence on Juan de Álava's desire to create rib designs that intertwine between the different segments. Keeping the star shape in each segment of the nave, the subsidiary ribs generate links which confer unity and flow on the entire nave; the result is the kind of web design found so frequently in the vaults of central Europe. The perspective section in fig. 15 shows the nave's ridge line in transverse, which is a consequence of the architect's desire to standardize the curvatures of the arches. One can see, in fig. 15, that the central boss is located at the top of the semi-circular diagonal arch, indicated by the dotted line.

Fig. 15. Convent of San Esteban, Salamanca. A perspective section of a vault through its transverse ridge line makes it possible to see clearly the curvature of the ridge line and how the central boss stone is located at the highest point of the semi-circular diagonal rib

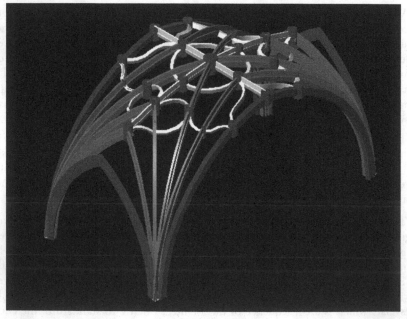

Fig. 16. Convent of San Esteban, Salamanca. 3D model of the spatial grid of ribs that compose the vault of a single bay

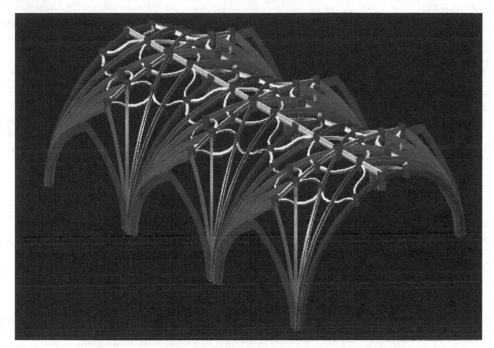

Fig. 17. Convent of San Esteban, Salamanca. 3D reconstruction of the spatial grid of ribs of the nave, probably the most spectacular vault ever built in Spain

## Cathedral of Plasencia, Cáceres. Presbytery vault. Started in 1522

The vault covering the presbytery of Plasencia Cathedral [Araujo 1997; Castro 2002; Navascués 1983] is also very spectacular (fig. 18). The front is formed with two vaults, one in the presbytery and one in the apse, intertwined by a daring design of subsidiary ribs based on concentric circles. In linking with one another, the secondary crossings generate a very elegant waving movement.

Fig. 18. Cathedral of Plasencia, Cáceres. Vault over the presbitery, Juan de Álava, started 1522

In order to achieve that effect, Álava thins and standardizes the sections of the arches as much as possible, especially the transverse arch which, because it is practically equal to the others, prevents the vault from being fragmented in segments.

Besides, with its 17.30 m span, this is one of the biggest Gothic vaults built in Spain. The width of the bay studied (7.34 m) means that its plan rectangle is remarkably elongated, exceeding the proportion 2 : 1. No doubt a division into such long rectangular bays has important structural consequences since, because the bays are narrower, there is an increase in the number of transverse arches and, on the outside, in the number of buttresses. Álava's great skill is evident when he adopts this solid structural layout to emphasize the lightness and elegance of the vaults [Rabasa 2000].

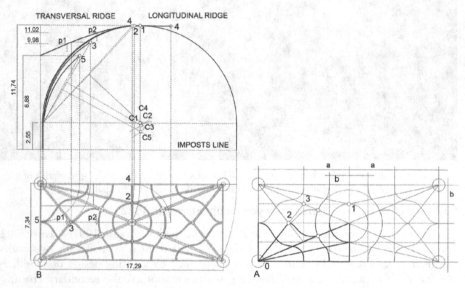

Fig. 19. Cathedral of Plasencia, Cáceres. Drawing of the vault located over the presbytery: A) plan; B) elevation of its arches. Notice the coincident curvature between the tierceron rib (2) and the transvers arch with the semicircular curvature of the diagonal rib. The curvatures of tierceron rib (3) and the diagonal have been made identical curvature by tilting them forward

As far as its morphological design is concerned, one can easily see that a 4x4 grid determines the locations of the most important bosses (fig. 19A). When looking at the four segments in which the small side can be divided, we realize that the first segment fixes the position of tierceron 1, the central circle's diameter and the links with the adjoining vault. The intersection with the vertical grid determines point 3, which in turn determines the position of tierceron 2. The remaining ribs seem to have been designed in a more random way.

In order to calculate the curvatures of the arches, we start by bringing up the diagonal (fig. 19b), which immediately shows that the two diagonal ribs are two extraordinarily stilted semi-circular circumferences, their curvature beginning 2.54 m above the capitals. This stilt contributes to create that impression of slenderness and lightness present in that vault.

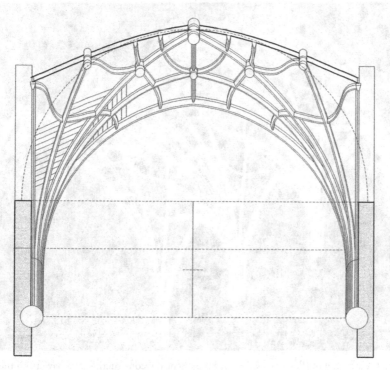

Fig. 20. Cathedral of Plasencia, Cáceres. Prospective cross-section showing the stilted springing as well as how the central boss stone is placed in the top of the semicircular diagonal rib (shown as a dotted line)

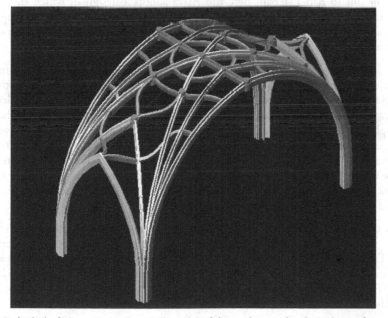

Fig. 21. Cathedral of Plasencia, Cáceres. 3D model of the vault over the choir drawn from the lines obtained geometrically. The image shows its pointed barrel vault shape

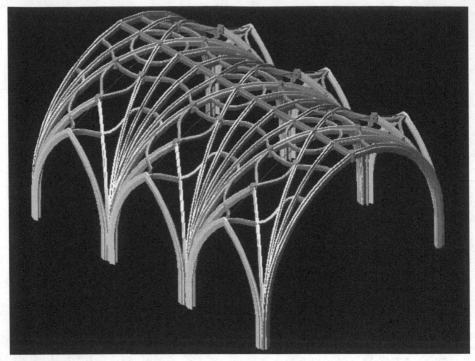

Fig. 22. Cathedral of Plasencia, Cáceres. Hypothetical reconstruction of a nave from the vault of the presbytery. This 3D model shows the intrincate grids of ribs that Juan de Álava was able to conceive

As he did in San Esteban, here too Álava adopts a design in which the transversal ridge line slopes steeply towards the formerets; in contrast, the ridge line along the nave axis is almost horizontal. As in previous cases, this configuration will result in creating a rather marked tunnel effect.

Unlike in San Esteban, in the Cathedral of Plasencia, when drawing the formeret and tierceron 3, according to the line of the transversal *rampant*, we see that neither of the two arches adjusts to the drawing of the diagonal semi-circular circumference; let us recall that each of the three arches is different. Nevertheless, bearing in mind the rational principles evident in Juan de Álava's work, we prefer to advance the hypothesis that the formeret as well as the tierceron 2 have been designed with the same curvature as the diagonal semi-circular circumference. If that hypothesis were correct, the centres of these arches would have to be below the impost level, and, therefore, the arches cannot be tangent to the vertical line. It is impossible to detect this feature visually, and it is further masked by the strong vertical stilt.

In the other direction the problem is easier, since a very slight slope of the rampant locates the bosses of tierceron 2 and the transverse arch in the curve of the semi-circular arch. The transverse arch assumes the form of an almost semi-circular pointed arch, with its centre placed next to its vertical axis, more or less at an eleventh part of the half of the span. Thus, if the hypothesis previously presented was correct, the whole of the vault could be carried out with just one arch.

Finally, a spectacular design of subsidiary ribs based on two concentric circumferences and some helpful ogees establishing four linking points with the apse's vault contribute to the creation of the waving movement of this rib work. The cells, as it is typical in Alava, are bonded *à la française*. The section in perspective in fig. 20 showse the rampant line and how the vaults rise on vertical stilts, most probably to allow them to reach the height of the vault of the transept.

The modeling of the vault of the presbytery in the Cathedral of Plasencia (fig. 21) again reveals Juan de Álava's mastery in conceiving and, later, building these three-dimensional webs. Fig. 22 shows how this rib work could have extended along the nave; although never built, it reveals the singular grid frames present in Juan de Álava's imagination.

## 7 Lierne vaults

It could be said that the main rib of the Gothic ribbed vault is the lierne, since its ridge line determines the height of the bosses which are situated on it. Curiously, in English Gothic this rib is always strictly horizontal, which places all the bosses along its run at the same height; as a consequence, the curvatures of the ribs with which the vault is built are subject to that inevitable starting condition. The visual effect of these sequences of bosses, in both the longitudinal and the transversal direction, is one of the most characteristic features of English Gothic style.

When English vaults accumulate tiercerons and countertiercerons, the conditional of the horizontal lierne results in a large collection of unequal arches. In order to rationalize the number of different arches, English Gothic architects developed the pointed arch with four centers known as the "Tudor arch".

There is, therefore, a significant difference between the "*flat rampant*" vault and the vault in which the lierne line is strictly horizontal. In the former, the diagonal arch has a semi-circular circumference, and the tiercerons are segments of this arch, that is, they are pointed arches with one center, whereas in the latter, the diagonal arch has three centers, which means that it is an oval, and the tiercerons are formed with this same oval, maintaining the springing arch and sliding the upper part of the oval onto the lower part in such a way as to keep the ridge line of the vault strictly horizontal. Thus, it is not an oval vault, since the rampants in it are not round.

Although the term "lierne vault" has a broader meaning in England, we have adopted this denomination for this particular type of vault and distinguish it from the "*flat rampant*" vault.

### San Esteban Monastery, Salamanca. Cloisters vaults. Built around 1533

Lierne vaults are apparently simple, but have a layout which conceals surprising features. One example is the vault in the Convent of San Esteban, Salamanca (fig. 23).

The cloister vault in San Esteban is actually a square vault with five bosses. The two tiercerons (see fig. 24A) are located in the position of the diagonal, between the transept and the formeret arch; we know perfectly well how to establish the position of this keystone through the circumscribed circumference and the straight line 1, 0 which, when crossing with the axis, determines point 2. In the center of the vault is a subsidiary rib whose diameter is set by the length 2-3, that is, the distance which separates the boss of the tierceron and that of the formeret. In each of its corners is a subsidiary rib that is one quarter of a circle, whose measure is determined by the alignment A, B, C; this is an open subsidiary rib, that is, one whose purpose is to connect with that of the adjacent bay in order to produce the linking of the ribs that is so characteristic of this architect's work.

Fig. 23. Convent of San Esteban, Salamanca, vaults in the cloister by Juan de Álava

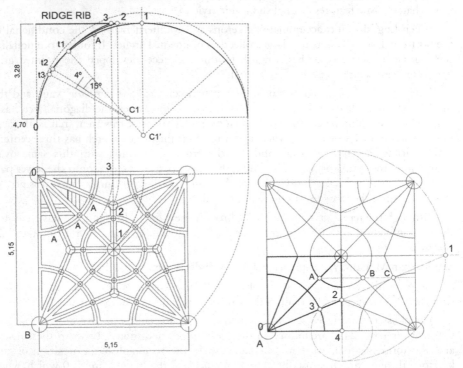

Fig. 24. Convent of San Esteban, Salamanca, cloister. Drawings of one of its vault: A) plan; B) elevation of the curvatures of its ribs. Observe that the diagonal arch has three centers and that the ridge line is a horizontal straight line. The other ribs are obtained by sliding the upper part of the oval diagonal onto the lower part, so that the curvatures of all ribs are identical

When laying out the arches (fig. 24B), we immediately note that the boss of the diagonal arch falls below the theoretical semi-circular arch; this is therefore a surbased arch which has to be laid out with three centers. Using the measurements taken on site, we have been able to draw the oval 1 and centers C1 and C1', which appear to fit the actual built arch perfectly.

We then proceed to draw the tiercerons, bearing in mind that they must reach the height determined by the vault's horizontal rampant. Obviously, the tierceron and the formeret could be pointed arches different from each other, but it is more advantageous to build them with the same diagonal arch, and thus to succeed in having the three arches making up this vault with the same layouts. To perform that, it suffices to rotate the upper section of arch 1 15° onto its lower section, so that boss 1 moves onto position 2, and the tangency point of oval 1 passes to 2. The boss of formeret 3 is carried out the same way, rotating the upper part of oval 4 onto the lower part. Proceeding this way, one obtains three different arches, which, however, have in common the upper and lower part; needless to say, this process resulted in significant simplification in cutting the voussoirs and the making the centerings.

The vault, being surbased, has a flattened ridge, which makes it possible to place the central wheel of subsidiary ribs at its summit, its circumference broken into few segments. This is also the case of the corner rib which, thanks to the fact that the three vaults arches are made from the same oval, can describe a continuous and horizontal circumference. The panels of the cells of the vault are bonded *á la française*. Fig. 26 shows some of the most outstanding circumstances of the vault: its horizontal rampant, the location of the boss below the semi-circular arch and its groined panel-work. In the three-dimensional image of a section (fig. 26), one can clearly see the volume obtained.

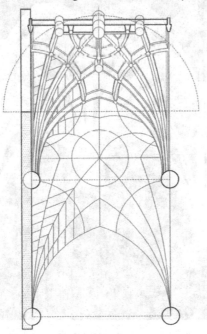

Fig. 25. Convent of San Esteban, Salamanca, cloister. Prospective cross section of the vault showing its horizontal ridge line and the position of its central boss stone. The stone masonry of its web is placed *à la francaise*

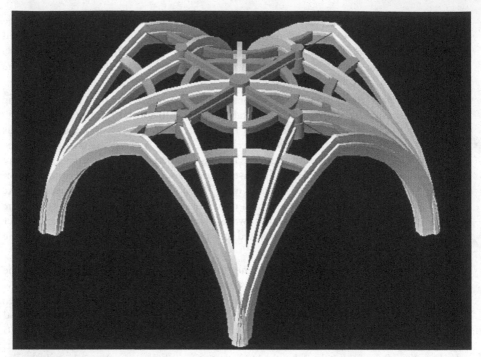

Fig. 26. Convent of San Esteban, Salamaca, cloister. 3D drawing of the resulting vault

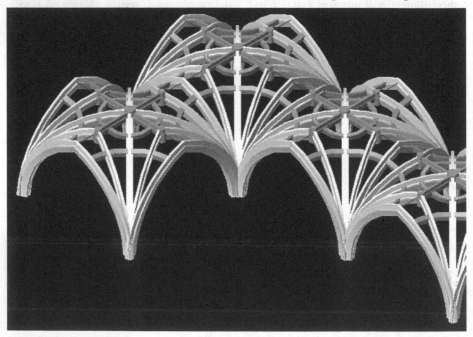

Fig. 27. Convent of San Esteban, Salamaca, cloister. Spectacular assembly of ribbed vault in the corner of the cloister

We previously noted that one of Juan de Álava's outstanding characteristics is his special way of laying out the subsidiary ribs so that they link the vaults of adjacent bays to generate an impression of continuity; this effect is enhanced when the section of the transverse arch is reduced to make it similar to the rest. Fig. 27 allows us to see the effect obtained with this type of subsidiary ribs and visualize the trumpet bell shapes of ribs created by this type of vaults from the springing to the ridge line lierne. This is particularly remarkable when the vault turns the corner in the cloister, where the springings of three of these vaults converge.

## 8 Flat Vaults

Flat vaults were very popular in Spain, more more than in the surrounding European countries, due to the emergence of a very particular typology of convent churches in Spain, in which the choir is situated on a high gallery over the first bays of the nave. Let us recall that, whereas in most European churches the choir was located in the presbytery, in Spanish cathedrals the choir occupied the central part of the main nave; nevertheless, in convents, the choir has a new location at the far end of the central nave, on an elevated storey through which one enters the church. This generally large, horizontal space is built on top of some vaults which needed to be surbased in order not to make this level excessively high and permit the visual communication between this platform and the high altar.

The huge expansion of religious orders in fifteenth- and sixteenth-century Spain increased the development of this type of church, and, consequently, the use of surbased vaults built with rib work. As we will now see, the layout of ribs in these vaults can be approached in two different ways: either through the use of very surbased basket-handle arches, or the use of segmental arches. In both cases, we come across the objective of reaching the highest possible standardization and simplification of the number of arches.

### Convent of San Marcos, León. Vaults of the *Sotocoro*. Built between 1531 and 1538

In the first place let us see how to construct out a flat vault by using basket-handle arches, that is, three-centered arches or portions of these. One example of such vaults are those which support the impressive choir of San Marcos en León, built by Juan de Álava between 1531 and 1538 [Castro 2002] (fig. 28).

Fig. 28. Convent of San Marcos, León, flat vault under the elevated choir

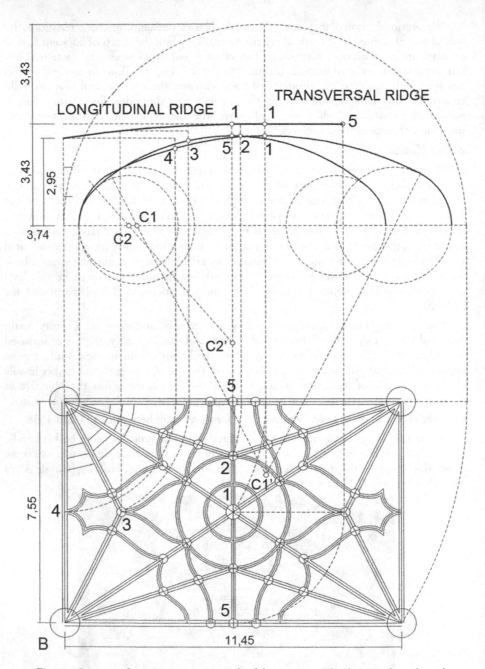

Fig. 29. Convent of San Marcos, León, vault of the *sotocoro*. The drawing shows how the curvatures of its ribs are obtained. All curvatures of the ribs, except the transversal arch, are identical to that of the three-centered diagonal arch. Once again, all ribs have been designed with the same curvature

Fig. 30. Convent of San Marcos, León, vault of the *sotocoro*.
Prospective section showing the surbased profile of the vault obtained with the oval diagonal rib,
exactly a quarter of the span of the vault

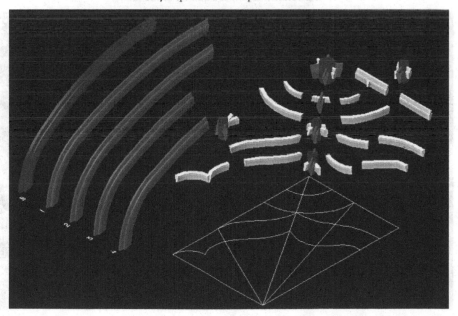

Fig. 31. Convent of San Marcos, León, vault of the *sotocoro*.
Standard parts required for the construction of this vault. All ribs, except the transversal arch, are
obtained by cutting the diagonal rib

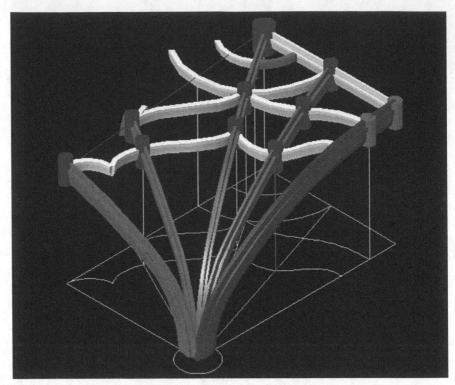

Fig. 32. Convent of San Marcos, León, vault of the *sotocoro*.
3D reconstruction of a quarter of the vault

Fig. 33. Convent of San Marcos, León, vault of the *sotocoro*.
Volumetric view of two bays of the *sotocoro* showing the intricate grid of ribs

This is a rectangular plan vault which is due to a strict modulation; the sides of the rectangle are in the ratio of 3:2; we are thus dealing with a *sesquiáltera* proportion. Its sophisticated design responds to a regular grid of 6x4 modules, which makes it possible to situate the most important bosses, while the others can be located by alignments. This rib work serves two aesthetic purposes: on one hand, a central design is established, underlined by double circles; on the other, the subsidiary ribs create another set of ribs which, terminating abruptly in the transverse arch, will create a continuous web interlinking the vaults. We have already noted Juan de Álava's particular way of carrying out the rib work.

Beginning to draw the curvatures of the arches (fig. 29), let us drop the diagonal and draw a semicircular arch in elevation as a reference. We will, then, locate the central boss in its precise height, and we immediately verify that this boss is located at the midpoint of the radius, so the transverse arch will be half as high as the semicircular arch.

In this case, the springings of the transverse arches are located at the impost level, so the axis of the oval described by this arch maintains a 2:1 ratio. The oval of the diagonal has been drawn with the centres C1 and C1', obtained by approximation. This basket-handle arch will be the one that generates the whole vault.

From the point 1, the central boss, can be drawn the two ridge lines of the vault. The smaller axis is a straight and horizontal line, whereas the rampant of the upper axis is a line, of a light curvature, descending with a slight slope. These two lines determine the height of the rest of the bosses.

When drawing the tiercerons 2, 3 and the formeret 4, we verify that their bosses are located precisely on the diagonal arch, which means that the three arches are portions of this arch. Therefore, the vault is designed in such a way that the layout of each one of them is obtained by cutting the diagonal arch with the span of the various arches. We thus find four equal arches; needless to say, this shows the obvious desire of the architect to rationalize the processes of drawing, cutting and centering of this sophisticated rib work.

It could be argued that, if the rampant in the longitudinal direction of the nave is horizontal, the tierceron 2 cannot be, strictly speaking, a portion of the diagonal arch; nevertheless, since the diagonal is such a surbased arch, its top remains practically horizontal for quite a long distance, which is sufficient to carry out the cut with which is obtained tierceron 2. In any case, it could be done rotating the arch's upper part onto the lower part, as we noticed when talking about lierne vaults.

The transverse arch, on the contrary, is different; its boss is at the same height as the central boss and its span is that of the vault's larger side, so this arch requires a specific oval. The basket-handle arch drawn with the centres C2 and C2' seems to adjust to these requirements.

Thus, such an apparently complex vault can be built with only two different arches: the diagonal and the transverse arch. Fig. 30 shows a perspective section in which one can observe the gradient of the longitudinal *rampant*, the line derived from progressively cutting the diagonal arch with the span of tierceron 3 and the formeret 4. In the figure, we can also see the panels laid in horizontal courses, "in the English way".

As shown in Fig. 31, we can envision the set of pieces required to carry out the vault: first, the collection of arches which, except the transverse arch, are clearly portions of the diagonal arch, that is to say equally traced. Then, one can see the six different bosses required by the vault and the set of subsidiary ribs which are fragments of flat ribs. As can be seen in fig. 32, with this repertory of pieces a quarter of the vault could be built.

Fig. 33 is a model of this vault in which we can well appreciate the resulting volumetry and its complex rib work. The image shows the spatial grid formed by the ribs of the three vaults making up the *sotocoro*, all of them built with arches generated from the diagonal's oval. It also allows us to verify the great skill of Gothic architects in the sixteenth century in solving these imposing three-dimensional webs.

### San Esteban Convent, Salamanca. *Sotocoro* vaults. Built between 1524 and 1610

The vault of the *sotocoro* in the convent of San Esteban de Salamanca (fig. 34) was built by Juan de Álava. It is a very surbased flat vault obtained by using oval arches of a very low rise. It has a rectangular plan measuring approximately 15 x 7.5 m; fig. 35 shows its complex plan and the ridge of its two sections. For the longitudinal rampant (in the direction of the nave axis), Juan de Álava decided on a straight, horizontal line, which means both the central boss and that of the transverse arch must be at the same height, but in the transversal direction he fixed a slightly curved rampant with a drop towards the boss of the formeret arch. These two lines are essential, since they precisely define the shape of the vault as well as determine the height of the secondary bosses (fig. 36).

Fig. 34. Convent of San Esteban, Salamanca.
Flat vault of the sotocoro built by Juan de Álava

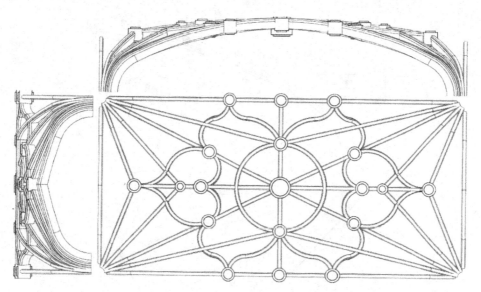

Fig. 35. Convent of San Esteban, Salamanca, flat vault of the *sotocoro*. Plan and sections of the vault showing its horizontal longitudinal section and its cross section curved and inclined over the formeret arches

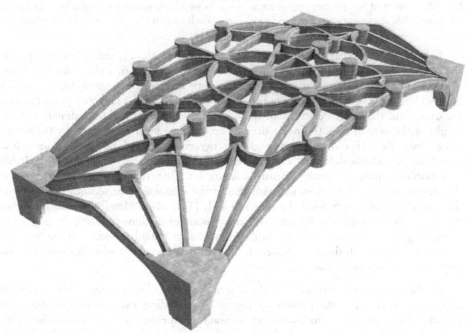

Fig. 36. Convent of San Esteban, Salamanca. 3D reconstruction of the vault of the high choir of the (drawing by Jorge Cerdá)

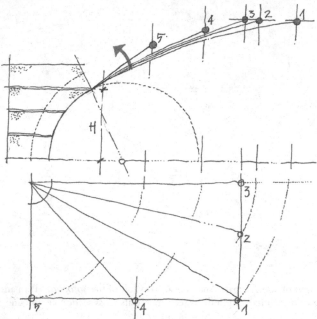

Fig. 37. Convent of San Esteban, Salamanca. Sketch in which the strategy followed by the architect to standardize the curvatures of the arches is shown. Each vault is laid out using the upper part of the diagonal oval arch

In order to fix the curvature of the tiercerons, Juan de Álava seems to have resorted to the strategy summed up in the sketch shown in fig. 37. In the first place, the lower part of the oval, which is not a real arch, coincides with the *tas-de-charge* of the vault and is built with horizontal beds. These courses rise to the height of the tangency point between the upper and the lower curvature of the diagonal oval, which had been drawn with the height of the central boss (point 1) deemed appropriate by the architect. In the second place, Juan de Álava decided that all the tiercerons should be oval arches whose curvatures are identical to that of the diagonal arch. Then, two conditions which will remarkably simplify the vault's construction are set. First, the lower part of the arch of the tiercerons is made to coincide with that of the diagonal arch; second, the height of the *tas-de-charge*, which fixed the diagonal arch H, is made identical in all the arches. Subject to these conditions, the various tiercerons are easily drawn, rolling the upper part of the diagonal arch from point H until it coincides, at the other end, with the height of each of the bosses. All things considered, the vault has been built with just one arch: the upper part of the oval of the diagonal rib.

At the Escuela de Arquitectura de Madrid, we had the opportunity to verify the hypothesis proposed in the previous paragraphs by building a scale model of this vault; the experience allows us to assert the accuracy of the precedent considerations. We proceeded first to lay it out at full-scale in plan and elevation (fig. 38). Following the patterns that we know of medieval stonecutting workshops, we extracted the data necessary to carve of each of the pieces that make up the vault (fig. 39). Once the carving process finished, we proceeded to the assembly of the vault, placing the pieces previously carved on a wooden centering (fig. 39). It is obvious that the centerings, all of a single curvature, are extremely simplified by the information provided. The final result is a scale model of the vault confirming the proposed hypothesis (fig. 40).

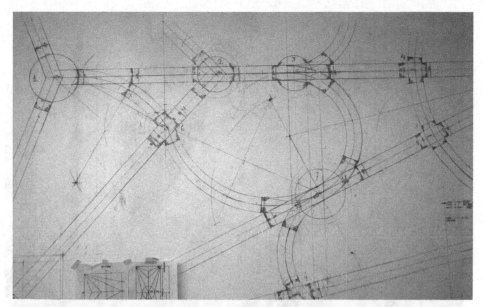

Fig. 38. Full-scale tracings drawn on the wall destined for the construction of a reduced-scale model of the vault

Fig. 39. Set-up; the different pieces composing the vault are placed on the centerings. The vault was built with only one radius of arch

Fig. 40. Reproduction of the sophisticated vault by Juan de Álava. This model was built at the School of Architecture of Madrid at a scale of 1:3, in the 2010 academic year

## 9 Conclusion

From its origins, the Gothic vault was conceived as an arrangement of cells whose folds are curved lines materialized in stone ribs. Thus, the Gothic vault of complex rib work can be described as a set of interconnected curved lines which forms a spatial grid; the vault is then formed when a thin shell of panels is spread over that grid. In short, Gothic stonemasonry relied on the definition of lines, and, to that purpose, created a geometric instrument that made it possible to carry out that task. It was with that goal in mind, and at that particular moment, when, during the long winters of inactivity and at the drawing boards, were generated the rudiments of the projection system nowadays known as dihedral, that is, a geometric method for establishing exactly the concordance between the plan and the elevation.

One result of that idea is the notion that there is no form of Gothic vault established a priori, but rather its form results from the spatial grid we just referred to. The grid's shape is a result of placing a series of points – the bosses – in space; these points are then connected by arches which are sometimes drawn from the vault's springing and sometimes from other bosses.

Gothic art is subject to a degree of increasing complexity; the subdivision of any of its elements has an influence on its vaults and the system is capable of drawing constructive lessons from that aesthetic drive. The increasing fragmentation of the panel cells results in their being easier to build; it is interesting to note how the multiplication of ribs is carried out according to very intelligent criteria for homogeneity and standardization.

In contrast, the Renaissance vault relied on perfectly defined form and was adjusted to pure prismatic shapes, always perfect from the geometric point of view, whether they are surfaces of revolution, toruses, ruled or prismatic surfaces. However, the breakdown of these surfaces into voussoirs results in blocks of great volumetric complexity requiring a more developed geometry, one which, though based on the same foundations as the Gothic, goes beyond that in order to understand and define very complex volumes, such as the voussoir work created in Renaissance vaults. From that time, on one can talk of stereotomy, that is, the cut (*tomos*) of the space (*estéreo*).

We can, thus, sum up by saying that the stonemasonry of the Gothic vault in its totality is based upon geometry of the line, whereas Renaissance stereotomy relies on the comprehensive knowledge of the surface and the highly sophisticated surfaces of the voussoirs necessary for its vaults. But it is obvious that this leap in the art of construction was paralleled and accompanied by an extension of the horizons of geometry. In Spain, it could be carried out thanks to the centuries-old tradition of building in stone that began in the most remote medieval times, and to the presence of architects and stonemasons as outstanding as Juan de Álava, whose professional work exceeded the established limits and provided the art of building with new instruments.

## Acknowledgments

This article is included in the research project "Stonecutting technology in the Mediterranean and Atlantic areas. Survey and analysis of built examples" (BIA2009-14350-C02-02)", sponsored by the Ministry of Science and Innovation of the Spanish Government in the context of the National Plan for Research, Development and Innovation.

## References

ARAUJO, Sebastián and Jaime NADAL. 1997. *Las catedrales de Plasencia, Historia de una restauración*. Cáceres : Colegio Oficial de Arquitectos de Extremadura.

BECHMANN, Roland. 1981. *Les racins des cathédrales*. Paris: Payot.

———. 1991. *Villard de Honnecourt, la pensée technique au XIII siècle et sa communication*. Paris: Picard.

CASTRO SANTAMARÍA, Ana. 2002. *Juan de Álava, arquitecto del Renacimiento*. Salamanca: CajaDuero.

CALVO LÓPEZ, José. 2009. La literatura de la cantería: una visión sintética. Pp. 101-156 in *El arte de la piedra*. Madrid: CEU Ediciones.

CHOISY, Auguste. 1899. *Histoire de l'architecture*. 2 vols. Paris: Gauthier-Villar.

CHUECA GOITIA, Fernando. 2001. *Historia de la arquitectura española*. 2 vols. Ávila-Madrid: Fundación Cultural Sta. Teresa: C.O.A.M Ávila.

FRANKL, Paul. 1962. *Gothic Architecture*. Harmondsworth: Pelican Books.

GARCÍA, Simón. 1991. *Compendio de arquitectura y simetría de los templos* (1681). Introductions by de Antonio Bonet Correa and Carlos Chanfón Olmos. Valladolid: Colegio Oficial de Arquitectos de Valladolid.

GÓMEZ MARTÍNEZ, Javier. 1998. *El gótico español en la Edad Moderna: bóvedas de crucería*. Valladolid: Universidad.

MERINO DE CÁCERES, José Miguel. 1996. *El claustro de la catedral de Segovia*. Segovia: Diputación Provincial.

MÜLLER, Werner. 1990. *Grundlagen gotischer bautechnik: ars sine scientia nihil est*. Munich: Deutscher Kunstverlag.

NAVASCUÉS PALACIO, Pedro. 1984. *Monasterios de España*. Madrid: Espasa Calpe.

PALACIOS GONZALO, José Carlos. 2003. *Trazas y Cortes de Cantería en el Renacimiento Español*. Madrid: Munilla-Lería.

———. 2009. *La cantería medieval, la construcción de la bóveda gótica española*. Madrid: Munilla Lería.

———. 2010. "Descimbrado de la bóveda gótica de san Esteban". YouTube. Madrid: UPM.

PALACIOS, José Carlos and Rafael MARTÍN. 2009. La construcción de una bóveda de crucería en la Escuela Técnica Superior de Arquitectura de Madrid. *Informes de la Construcción* **61**, 515: 49-58.

RABASA DIAZ, Enrique. 2000. *Forma y Construcción en piedra, de la cantería medieval a la estereotomía del siglo XIX.* Madrid: Akal.

———. 2005. Construcción de una bóveda de crucería en el centro de oficios de León. Pp. 909-917 in *Actas del cuarto Congreso de Historia de la Construcción,* Cádiz, vol II. Madrid: Instituto Juan de Herrera.

VIOLLET-LE-DUC, Emmanuel. 1854-1868. *Dictionnaire raisonné de l'architecture française du XI au XVI siècle,* 10 vols. Paris: B. Banche Editeur.

———. 1878. *Histoire d'une hôtel de ville et d'une cathédrale.* Paris: J. Hetzel. Rpt. Bruxelles: Pierre Mardaga, 1978.

WENDLAND, David. 2009. *Lassaulx und der Gewölbebau mit selbsttragenden Mauerschichten. Neumittelalterliche Architektur um 1825-1848.* Petersberg: Imhof.

Willis, Robert. 1842. *On the constructions of the Vaults of the Middle Ages.* London. Rpt. Transactions of the RIBA, Longman, Brown, Green and Longmans, 1910.

**About the authors**

José Carlos Palacios Gonzalo received his Ph.D. in Architecture in 1987. He is Senior Lecturer at the Department of Construction of the Escuela Técnica Superior de Arquitectura in Madrid. Universidad Politécnica de Madrid (UPM). Lecturer in "Gothic Construction Workshop". This subject won the award to Innovation in Education 2009 of the UPM. For eleven years he has been Visiting Professor in the Master in Restoration at the Faculty of Architecture in Louvain (Belgium), giving a course on stereotomy. He is the author of the book *Trazas y cortes de cantería del renacimiento español* published in 1990 (second ed. 2003). He has also written *La cantería medieval. La construcción de la bóveda gótica española* (published in 2009).

Rafael Martín Talaverano received his degree in Architecture in 2002 from the Universidad Politécnica de Madrid, and a Master in Architectural Restoration in 2004 from the same university. He is an instructor at the School of Architecture of Madrid in Gothic Construction Workshop, Constructive Systems 2007-2009, Descriptive Geometry, and Computer Graphics Workshop from 2010 to the present. He is an Associate Professor. He has published a number of papers on construction history in international conferences and refereed journals.

# Ana López Mozo

Escuela Técnica Superior de
Arquitectura
Universidad Politécnica de
Madrid
Av. Juan de Herrera 4
28040 Madrid SPAIN
ana.lopez.mozo@upm.es

Keywords: ovals, vaults, arches,
history of construction, the
Escorial, Juan de Herrera

Research

# Ovals for Any Given Proportion in Architecture: A Layout Possibly Known in the Sixteenth Century

**Abstract.** Oval forms have been used in architecture since antiquity as arch elevations, cross and horizontal sections of vaults, profiles of arches and building plan layouts. The present paper aims to approach the knowledge and application of oval layouts for any given proportion, that is, those which fit a particular place, either as the span and height of an arch or vault, or as the length and width of a plan, by comparing written sources and built heritage. This is, on one hand, research on architectural treatises since the sixteenth century in order to find where the geometrical construction for this kind of layouts appears for the first time; and on the other, a study of the application of ovals in the vaults of the Escorial (1563-1584). Although 1712 could be considered the date of the first published geometrical construction for an oval to fit a given place, this work hypothesizes the possible application of a layout of this type in the construction of the Escorial.

## Introduction

An oval is a closed convex planar curve, with double orthogonal symmetry, made of at least four tangent circular arcs. The figure can be drawn using a compass and a ruler. In the words of Spanish architect Fray Lorenzo de San Nicolás, "the oval is an elongated circular figure" (my translation) [1639: 149]. Ovals and ellipses show a clear graphical similarity, but are quite different in terms of their generation, layout and drawing processes, as well as the implications concerning their architectural construction. In fact, the use of ovals in building is much more common [Huerta 2007: 211; López Mozo 2009: 508]. The word oval has been used, however, to refer to both oval and elliptical layouts [Gentil 1996: 83-84]. Juan Caramuel de Lobkowitz was the first among the authors of architectural treatises to apply the names of oval and ellipse correctly [1678 Treatise IV: 28, Fernández Gómez 1994: 349]; Caramuel described the oval as an imperfect ellipse. Although the use of oval forms in architecture comes from antiquity, starting in the Middle Ages oval domes were frequently used to cover rectangular rooms and, in particular, surbased arches and vaults were constructed in height-limited spaces [Palacios Gonzalo 1990, 2003; López Mozo 2000, 2003, 2005; Huerta 2007; Martín Talaverano 2007, 2009].

A comparison of written sources and built heritage shows that the two sides of reality, theory and practice, do not always concur. The study of the oval layouts found in the treatises confirms the state of the knowledge at a certain moment; on the other hand, the analysis of the ovals actually built at the Escorial gives us data on architectural building practice. As we will see, oval layouts in the Escorial were constructed to fit any measure by trial-and-error adjustment, as had been usual since the late Gothic, when the use of surbased vaults spread [Huerta 2007: 222].

The present paper contains the following parts: a short discussion of the geometrical properties of ovals and ellipses and the surveying criteria to determine the nature of an oval form actually built; a research on oval layouts for any given proportion in architectural treatises since the sixteenth century; an analysis of the application of ovals in the stone vaults of the Escorial and a description of the oval layout for any given proportion that could have been used there.

## Ovals and ellipses: Geometry and surveying criteria

Although the discussion *oval versus ellipse* is not the main focus of the present paper, it is necessary to begin by defining geometrical differences and their consequences in architectural application. As noted, an oval is a figure made of tangent circular arcs. The condition of tangency only demands an alignment of the two centres of the consecutive arcs with the point of change of curvature. The layout of a complete symmetrical oval requires four circular arcs and their corresponding centres, but this number may be increased. Drawing parallel ovals in order to generate walls or arches with a constant thickness or surrounding aisles of a given width is simple as all of them are concentric circular arcs. It may be noticed that proportion between the two axes of these parallel ovals is not a constant number. The layout of orthogonal lines on an oval in order to locate, for instance, the bed joints of the voussoirs of an arch is easy, as they are aligned with the centres of the arcs. As we will see, the drawing of ovals generated from a previous geometrical construction historically began with the layout of oval figures of some standard proportions, which do not fit any pair of axes: the first published proposals come from the sixteenth century.

The ellipse was known from antiquity, and its generation was explained geometrically as one of the three kinds of planar sections obtained from a cone; metrically, it was defined as the locus of the points of the plane whose sum of distances to two fixed points gives a constant result. Drawing an ellipse by the so-called 'gardener's method' applies the last definition and is useful in large drawings; at a small scale, however, the manual procedure is only an approximation, placing isolated points and requiring the final line to be drawn freehand.[1] In the mentioned *gardener's method* – also called the *string method* [San Nicolás 1639, Part I: 67r; Huerta 2007: 215] – the free length of the string, once it is tied, coincides with the dimensions of the major axis. In addition, half the value of this length is the distance from the endpoint of the minor axis to any of the foci, which can be easily located in this way. Bachot [1598] clearly explains the method by three figures [Huerta 2007: 234]. The curvature of the ellipse varies at all points, which complicates the division into equal parts and the layout of orthogonal lines to distribute the voussoirs of an arch and the implementation of instruments for dressing and checking. On the other hand, an equidistant line of an ellipse is not another ellipse but a fourth-degree curve called a *toroid* [Gentil 1996: 81; Huerta 2007: 217], which can only be drawn by approximation. It is thus complicated to lay out the parallel curves that usually form the extrados of walls and arches and width of surrounding areas.

Determining the nature of an oval form that is actually built means discarding the elliptical layout on one hand, and identifying the original oval layout on the other. If we are studying a set of parallel ovals we should identify the primary one, where geometrical constructions were applied. As noted, equidistant ovals do not maintain proportional properties: we will not find an equal division of the axis to place the centres or a certain ratio between maximum and minimum diameter if we are not working on the first designed oval of the set. Given the slight differences among all oval and elliptical options, these decisions are difficult to make, even if the measurements of the data are exhaustive and rigorous [Huerta 2007: 217]. The surveying process demands a specific control of both measuring and data analysis, which will now be described for a barrel vault.

There is a relationship between the deformation of a vault and the inclination of its walls – by means of the thrusts – which is larger in a wall with doors or windows and smaller if the vertical load coming from above is larger. Thus, if we want to approach the original layout of the vault, the profile should be measured in an area with sufficient counter-thrust; if walls are supporting the vault in the direction of the cross-section, this is the right place, as the deformation beside them will be almost zero. If we are measuring a barrel vault around a cloister, we should take points near the corners, where the inclination of the walls and the deformation of the vaults will be smaller.

The analysis of data from a surbased profile could be done according to the following steps. We should first draw the ellipse whose axes coincide with the main dimensions of the arch and compare the curve to the measured points. If the match is not satisfactory, then the layout corresponds to an oval; the next step will be to check the oval layouts. Even if the coincidence is fully satisfactory, we still cannot conclude that there is a layout of an elliptical nature, because it is possible to draw an oval fitting any measure by means of trial-and-error adjustments, even before the geometrical construction for that problem has occurred. So the next step will always be to check the oval layouts in order to either determine which kind of oval it is or to conclude that it is an ellipse. We should first compare the measured points to the existing layouts of fixed proportions, all along the arc. The coincidence of proportions between major and minor axes is not enough evidence of the application of an oval layout: as we have said, it is easy to fit different ovals into the same place. If the existing oval models do not coincide with the data of the vault, we should adjust an oval layout by the modern method: we first try applying a circular arc into the area over the springing line and drop the length of its radius from the key of the arch; we draw a straight line from that point to the centre of the circular arc and then its median line, which will meet the vertical axis in the centre of the large arc of the oval. By drawing a straight line from one centre to the other and extending it towards the lateral arc, we define the radius of the central arc, which can then be drawn. After this first approach to the layout, we should adjust the position of the centres and the size of the radius by comparison with the original measuring unit and by checking common construction procedures, such as the division of the span into equal parts, or placing the centres to form an angle of 30°, 45° or 60° to the horizontal line.

Despite all the precautions taken, it is usually necessary to fall back on original written or graphical sources or arguments of logical constructive practice to be able to determine the nature and original layout of an oval form.

## Oval layouts in the treatises since the sixteenth century

It is generally accepted that the geometrical construction for drawing an oval for any given proportion was not known in the sixteenth century. Although, mainly since the Middle Ages, masons used to build oval forms fitting any measure by a trial-and-error adjustment, it would be important to know where and when the geometrical solution to the problem was reached, and when the architects had more design tools at their disposal. Research of this kind of oval layouts in the treatises on architecture is shown below. The sixteenth century was considered to be a good starting point, since there we find the first published oval layouts. These Renaissance drawings have been carefully studied by scholars [Fernández Gómez 1994; Gentil 1996; Huerta 2007]. They do not fit any given proportion, but they are included here because they were copied by almost all subsequent authors, establishing a starting point and fundamental reference.

Serlio was the first to solve the problem of laying out a surbased arch of a certain height and span by an arc that is now called elliptical and explained by an affine transformation of the circumscribed and inscribed circumferences (radii equal to the half-span and height of the arch [1545: fol. 13-14]). The layout is easy to draw, as what is necessary is not to divide the circumferences into equal parts, but rather to draw any set of straight radial lines. As a result, the problem of laying out arches of any height was solved; the same cannot be said of the ovals of fixed proportions that Serlio shows afterwards [1545: fol. 17-20]. The author draws four different ovals that are in fact three layouts, as the first one is only a general description of the drawing of parallel ovals with their centres aligned so as to form a 60° angle to the horizontal. Each of them has a fixed proportion between its axes: √2, 1.3203 and 1.3227 [Gentil 1996: 84-90]. Therefore, none of Serlio's ovals can be adjusted to fit any given place, but only those which have the same proportion (fig. 1).

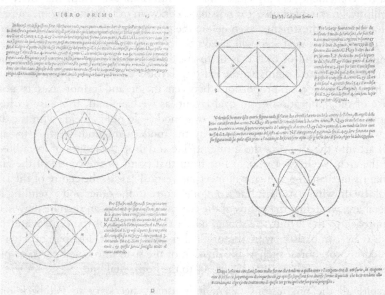

Fig.1 Four methods of drawing ovals from *Il Primo libro d'Architettura di Sebastiano Serlio* [1545]

The manuscript of the Spanish architect Hernán Ruiz shows a copy of the ovals by Serlio and includes a very interesting and unpublished layout, as it starts from the height of the arch, and not from the span [c. 1560: fol. 24v]. The drawing process is as follows: the centres of the large arcs are placed on the key of the arch and its symmetric point; two straight lines forming 30° to the vertical axis of the arch are drawn from those centres, determining in the horizontal axis the centres of the two minor arcs; the meeting point of the extended straight lines with the central arc determines the change of curvature, the radii of the minor arcs and, finally, the length of the major axis or span of the arch (fig. 2). The proportion between major and minor axes is 1.4226 [Gentil 1996: 90]. Hernán Ruiz also copies the surbased "elliptical" layout by Serlio and proposes two new ways to lay out surbased or stilted arches by stretching a circumference [c. 1560: 23, 37, 41v and 47v; Gentil 1996: 97; Huerta 2007: 225] (fig. 3).

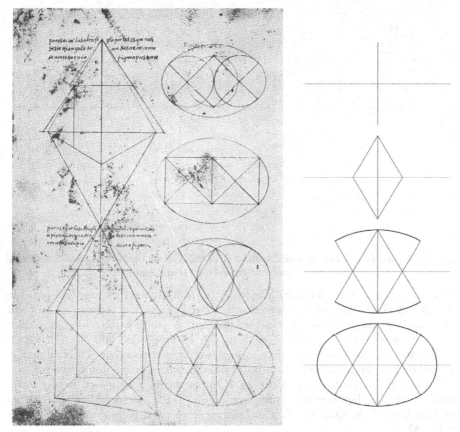

Fig. 2. Left, Hernán Ruiz, *Libro de Arquitectura*, c. 1560 [Navascués 1974]. On the right, possible drawing process of the oval below

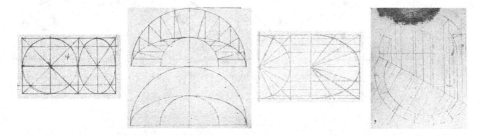

Fig. 3. Hernán Ruiz, *Libro de Arquitectura*, c. 1560 [Navascués 1974]

Philibert De l'Orme [1567: 118] draws a surprising oval cross-section for a surbased dome with a double proportion between its axes [Gentil 1996: 97-98]. It does not seem to follow a controlled geometrical construction: the span is not divided into equal parts to locate the centres of the lateral arcs, and the straight line connecting the change of curvature point with the centres of the consecutive arcs does not form a fixed angle 30°, 45°… to the horizontal springing line (fig. 4).

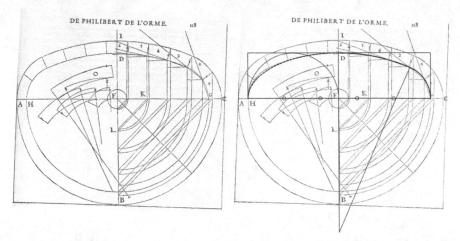

Fig. 4. De l'Orme, *Le premier tome de l'Architecture*, 1567. On the right, overlapped, an elliptical layout (dotted line) and an oval one (continuous line)

The stonecutting manuscript by Vandelvira [c. 1580] follows the storyline posed by Serlio: he describes only one oval layout with a fixed proportion between its axes and, then, in the next sheet of paper, copies the "elliptical" layout of Serlio, citing the author, to solve the problem of adapting an arch to a given proportion and "lay out a vault according to the demands of the place in question" (my translation). The oval arch or *carpanel*[2] by Vandelvira is lower than those by Serlio, with a proportion between axes of 1.5773. The geometrical layout of the arc is similar to the fourth oval by the Italian architect, with the alignment of centres forming 60° to the horizontal axis, but starting with a division of the span into four rather than three equal parts [Vandelvira c.1580: 18r] (fig. 5).

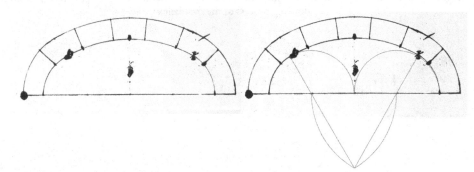

Fig. 5. Vandelvira, *Libro de Traças de cortes de piedras*, c. 1580.
On the right, overlapped, explicit geometrical layout

This is the only kind of oval layout used by Vandelvira in the entire treatise – which comprises 141 different stonecutting layouts – even though he knew the ovals by Serlio, which certainly give less surbased arches. Vandelvira's manuscript is a theoretical text, that is, he is free to design the measures and proportions of all the architectural models explained. His oval layout is simple and useful, as it can be drawn using only a compass. In the last page of the manuscript the author sets out an unpublished construction to draw an oval figure – we would now say "elliptical" – into a non-rectangular parallelogram. The procedure can now be explained as an affine transformation of a circumference [Vandelvira c.1580: 126r] (fig 6).

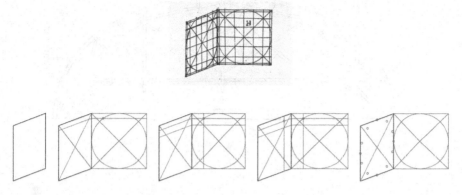

Fig. 6. Vandelvira, *Libro de Traças de cortes de piedras*, c. 1580, method of inscribing an "oval figure" within a non-rectangular parallelogram and graphical interpretation by the author

Cristóbal de Rojas [1598: 98] follows the idea of the layouts for arches by Vandelvira [c.1580: Tít. 21], but his drawing for the oval arch corresponds to the fourth layout by Serlio, with the alignment of the centres forming a 60° angle to the horizontal springing line, perhaps because it was clearly described graphically (fig. 7).

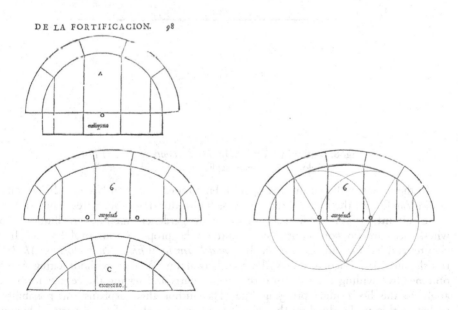

Fig. 7. Cristóbal de Rojas, *Teoría y Práctica de Fortificación*, 1598.
On the right, overlapped, explicit geometrical layout

The carpentry manual by Mathurin Jousse contains a drawing for an oval roof structure, with a layout that could follow the fourth oval by Serlio [(1627) 1702: 156] (fig. 8).

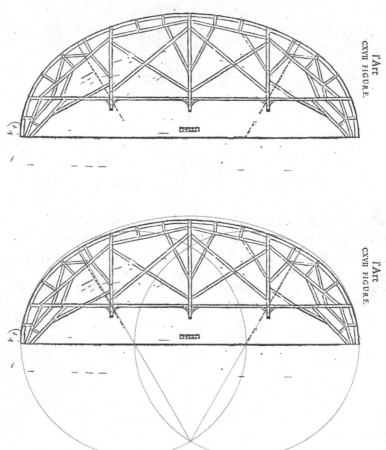

Fig. 8. Mathurin Jousse, *L'Art de Charpenterie*, (1627) 1702.
Below, overlapped, explicit geometrical layout

In the seventeenth century the Spanish Fray Lorenzo de San Nicolás [1639] follows, as did Vandelvira, the method set out by Serlio for drawing oval figures: when they have to be built into certain given measures he uses the gardener's method, an elliptical layout; when there are no constraints on proportions he proposes four oval layouts. In the chapter entitled *On Arches and the Way and Form in Which They Must Be Made* (my translation), San Nicolás draws the fourth oval by Serlio – without citing Serlio – obtained by dividing the span into three equal parts and having the centres form a 60° angle to the horizontal springing line. The author then explains the possibility of surbasing this arc by dividing the span into a larger number of equal parts, although he points out that a better way is the string method/gardener's method, with the advantage that it "can be surbased as it is desired" (my translation) [1639: fol. 66v-67v]. The drawing more closely approaches an oval than it does an ellipse, and the author, in fact, points out that it can be dressed with three *cintrels* (fig. 9).

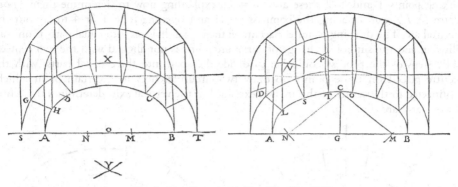

Fig. 9. Fray Lorenzo de San Nicolás, 1639

Later on, in the Chapter titled *On the Construction of Ovals and Their Measures and Other Advices* (my translation), San Nicolás describes three unpublished layouts and repeats the fourth oval by Serlio [1639: 150,151]. Following the order of his drawings, the first one is similar to the second oval by Serlio with a free placement of the centres, only keeping an equal distance to the centre of the oval. The process of drawing starts with the major axis: extending the straight line which connects the centres, he starts by drawing the small arcs, which define the radii of the large ones in the intersection with the mentioned line. The measure of the minor axis of the oval is obtained as a result of the drawing process. It could have also been done in reverse order, starting out from the minor axis and defining the major one. The second oval proposed by San Nicolás does not comply with the tangency conditions: the centres are not aligned with the change of curvature point. The third layout starts from a square that does not contain the final points of the axes, so the main measures of the oval are not previously controlled, but the construction is geometrically correct. Two of the centres are located at the midpoint of the square sides, and the remaining two can be found in the centres of the two rectangles obtained by the division of the square into two equal parts (fig 10).

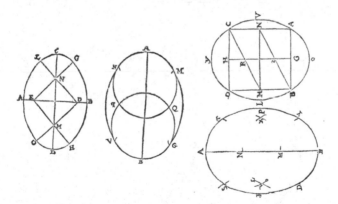

Fig. 10. Fray Lorenzo de San Nicolás, 1639

The stonecutting manuscript by Mathurin Jousse shows an odd oval layout [1642: 8]. Although the drawing process seems to be explicit both graphically and in written form, it is difficult to understand. The text proposes a division of the major axis into three equal parts, where the centres of the lateral arcs will be placed. Then the author suggests taking points 1 and 4 of these arcs, without explaining how to determine them, '*pour faire D, H, égal*', placing the compass on H and opening it to 1 or 4 to lay out the central arc. The drawing is quite inaccurate: the initial division into equal parts is only an illusion and the supposed change of curvature point is not aligned with the two centres. The steps of the process could be: to divide the span into three equal parts; with the centre placed on those points, to draw two circumferences to a certain point which, connected to the centre of the arc and extended to the vertical axis, determines the centre of the large arc (fig. 11).

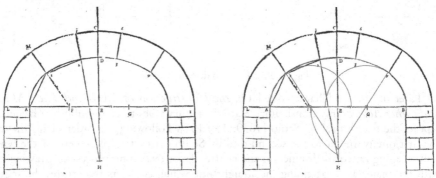

Fig. 11. Mathurin Jousse, *Le Secret d'Architecture*, 1642.
On the right, overlapped, explicit geometrical layout

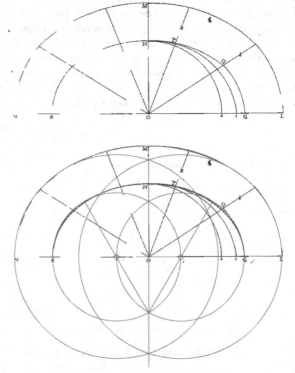

Fig. 12. François Derand, *L'Architecture des voutes*, 1643.
Below, overlapped, explicit geometrical layout

The stonecutting treatise of the French Jesuit François Derand [1643: 302] does not include explicit oval layouts but he does draw some. For instance, in the layout for a *Trompe en niche en demy-ovale, ou surbaisée, ayant mesme cintre que son plan*, the text describes an elliptical centring, but the drawing follows the fourth oval by Serlio (fig. 12).

The manuscript by the Majorcan architect Joseph Gelabert [1653: 21r] proposes a layout for an oval arch. Departing from the division of the span into five equal parts, the centre of the lateral arc is placed on the first point, and a straight line forming a 60° angle to the horizontal line is drawn, whose intersection with the vertical axis will determine the centre of the large arc [Rabasa 2011] (fig. 13).

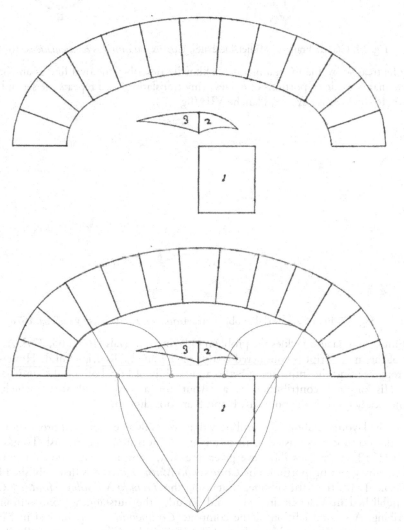

Fig. 13. Above, Joseph Gelabert, *De l'art de Picapedrer*, 1653; below, an oval layout after Rabasa [2011]

The French Jesuit Claude François Milliet-Dechales includes the second and fourth ovals by Serlio in his encyclopedia on mathematics [1674: Vol.2, 8] (fig. 14).

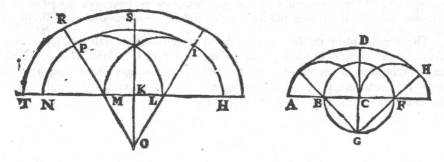

Fig. 14. Claude François Milliet-Dechales, *Cursus seu mundus mathematicus*, 1674

The treatise by Juan Caramuel de Lobkowitz describes the oval line as an "imperfect ellipse that is made of portions of circles" (my translation) and repeats the second oval by Serlio [1678: Treatise IV; 28, Planche VII] (fig. 15).

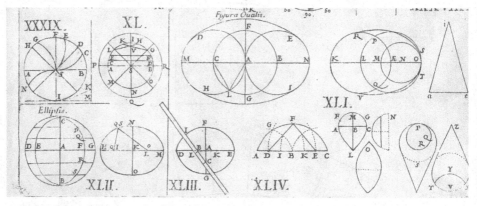

Fig. 15. Juan Caramuel de Lobkowitz, *Arquitectura civil recta y oblicua*, 1678

Simón García also tackles the problem of laying out ovals [1681: 65, 77v]. As the first part of the manuscript is considered a copy of the one by Rodrigo Gil de Hontañón,[3] in this topic we find the influence of Serlio and, mainly of Fray Lorenzo de San Nicolás (fig. 16). His original contribution is a layout for a semi-ovoid shape which is not geometrically correct: the points K, L and B are not aligned.

All the layouts analysed up until now draw ovals of a certain fixed proportion, that is, ovals that cannot be adjusted to fit to any given measures. The Spanish Tomás Vicente Tosca [1712] might possibly have given the first, general construction for drawing an oval for any given proportion. His *Compendio Mathematico* was first published between 1707 and 1715; the fifth volume, containing the *Tratado XV sobre Montea y Cantería,* was published in Valencia in 1712, and includes the outstanding contribution we are describing. A second edition of the complete *Compendio* was published in Madrid in 1727 and another one in Valencia in 1757 [Fernández Gómez, 1994: 349; 2000: 21]. Tosca's *Compendio* is believed to have been influenced by the work of Milliet-Dechales [Rosselló 2004: 160] and the text on stonecutting is said to be a copy of the French treatise [Rabasa 2000: 234, 306]. Tosca describes five ways to layout a surbased arch: two oval and three elliptical ones [1712, Treatise XV: 99-104 and 108]. The first oval method is the fourth proposed by Serlio and the second, which looks like a segmental arch, is the most interesting for the present research (fig. 17).

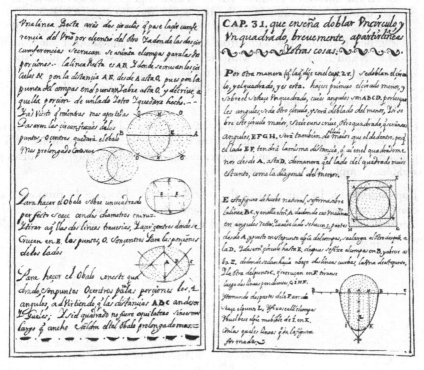

Fig. 16. Simón García, *Compendio de arquitectura y simetria de los templos*, 1681 [1990]

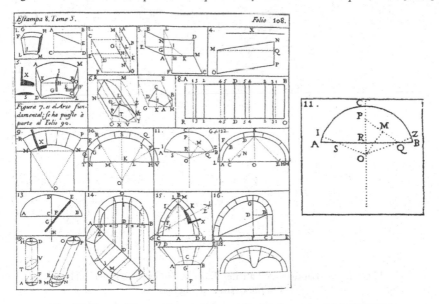

Fig. 17. Tomás Vicente Tosca, *Compendio Mathematico*, Vol. V [1712]

The translation of the text describing the way of drawing an oval fitting any given proportion is the following:

> Given the horizontal diameter AB of an arch (fig. 11) and CR the height that it must have: cut arbitrarily, but equal, the distances AS, BQ, CP. Draw the line PQ and its median line MO, which will meet to the CR lengthened to O. Draw a line from O to Q and Z; and another one from O to S and I; and from O with the distance OC draw the arc ICZ; and from Q with the distance QZ draw the arc ZB: and from S the arc IA, and the intrados will be formed: with the same centres the extrados will be drawn (my translation).

Indeed, choosing the radii of the lateral arcs of the oval (AS and BQ in the text of Tosca) fixes one of the infinite solutions to the problem. The importance of Tosca's contribution should be studied and analyzed taking into consideration the sources which he could have employed for this topic.[4]

Jean-Baptiste De La Rue ignored the contribution by Tosca, as he proposes a layout to draw a semi-oval given the diameter and the height, which, however, is not geometrically exact [1728: 5]. A graphical translation of the method by De La Rue is shown in fig. 18.

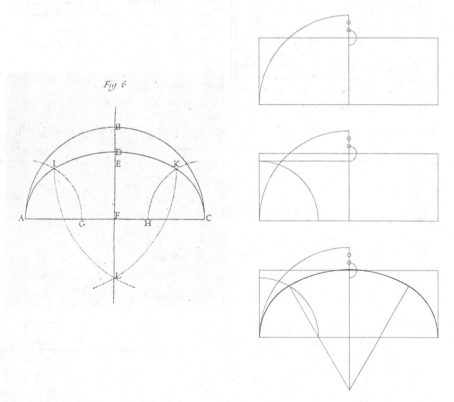

Fig. 18. Jean-Baptiste De La Rue, *Traité de la Coupe des Pierres* [1728].
On the right, drawing process

The mathematical proof of the erroneous construction by De La Rue is given below. As the centres are aligned forming 30° to the vertical axis of the oval, if the layout is correct, dropping the radius of the lateral arcs from the key and drawing a straight line to one of their centres should give an angle of 15 ° to the horizontal line. Letting $a$ and $b$ represent the lengths of the semi-axes, the proof would be:

$$tga = \frac{\frac{1}{3}(a-b)}{a-b+\frac{1}{3}(a-b)} = \frac{\frac{1}{3}(a-b)}{\frac{4}{3}(a-b)} = \frac{1}{4} = 0.25; \quad \alpha = 14.03° \neq 15°$$

The contribution by De La Rue is quite ingenious, as the drawing is very close to the exact one, especially in arches that are not excessively surbased.

Frézier, critical of the use of ovals [Huerta 2007: 234], poses the problem, "given two axes, imitate an ellipse by the union of four arcs of circumference" (my translation) and finds an interesting solution. Frézier describes an exact oval layout for any given proportion, with the alignment of the centres forming a 60° angle to the horizontal line, of which the mathematical proof is given but not completed [1737: 183, Pl. 14]. The drawing begins by taking the length of the minor semi-axis on $By$; the rest of the process is clearly explained in the figure (fig. 19).

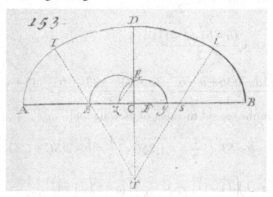

Fig. 19. Amédeé François Frézier, *La théorie et la pratique de la Coupe des Pierres*, 1737

The mathematical proof, for the oval starting from A and B and exactly reaching D, would be the next one:

$$AS + ST = DC + CT .$$

With $a$ and $b$ as the lengths of the semi-axes, we have

$$Cy = (a-b) \text{ and } CZ = \frac{1}{2}(a-b).$$

The first step should deduce the values of CE and SZ:

$$tga = \frac{CE}{a-b}$$

$$tga = \frac{\frac{1}{2}(a-b)}{CE}$$

$$\frac{CE}{a-b} = \frac{\frac{1}{2}(a-b)}{CE} \ ; \ CE^2 = \frac{1}{2}(a-b)^2 \ ; \ CE = \frac{a-b}{\sqrt{2}}$$

$$SZ = ZE = \sqrt{CZ^2 + CE^2} = \sqrt{\frac{(a-b)^2}{2^2} + \frac{(a-b)^2}{2}} = (a-b)\sqrt{\frac{3}{4}} = (a-b)\sqrt{3}\frac{1}{2}$$

Going back to the equation we have to prove, we will work separately with each of the members of the equation. Considering that the triangle $SST$ is equilateral, it must be hold true that $ST = 2SC$.

$$AS + ST = [a - (SZ + CZ)] + 2SC = a - (SZ + CZ) + 2(SZ + CZ) = a + SZ + CZ$$

Replacing $SZ$ and $CZ$ by the mathematical expression found above:

$$AS + ST = a + (a-b)\sqrt{3}\frac{1}{2} + \frac{(a-b)}{2}$$

$$AS + ST = a + \frac{(a-b)(\sqrt{3}+1)}{2}$$

$$= \frac{2a}{2} + \frac{\sqrt{3}a - \sqrt{3}b + a - b}{2} = \frac{3a + \sqrt{3}a - \sqrt{3}b - b}{2} = \frac{a(3+\sqrt{3}) - b(1+\sqrt{3})}{2}$$

If we operate in the second member of the equation:

$$DC + CT = b + ST\sqrt{3}\frac{1}{2} = b + 2SC\sqrt{3}\frac{1}{2} = b + 2(SC + CZ)\sqrt{3}\frac{1}{2}$$

$$= b + 2\left[\frac{(a-b)(\sqrt{3}+1)}{2}\right]\sqrt{3}\frac{1}{2} = b + \left[\frac{(a-b)(\sqrt{3}+1)}{2}\right]\sqrt{3}$$

$$= \frac{2b}{2} + \frac{(\sqrt{3}a - \sqrt{3}b + a - b)\sqrt{3}}{2} = \frac{3a + \sqrt{3}a - \sqrt{3}b - b}{2} = \frac{a(3+\sqrt{3}) - b(1+\sqrt{3})}{2}$$

Therefore, it has been proven that $AS + ST = DC + CT$.

The oval by Frézier is based on an exact geometrical construction, offering a unique solution to the problem: an oval that fits the given measures of the axes, with its centres aligned forming a 60° angle to the horizontal line. Like De La Rue, Frézier ignores the contribution by the Spaniard Tosca, who had found the most complete solution twenty-five years earlier.

The book *Escuela de Arquitectura Civil* by the Spanish architect Brizguz y Bru [1738: 16-17], published in Valencia like Tosca's *Compendio*, shows four oval layouts. The first three start out from the major diameter and use a similar method; two equal circumferences are drawn, passing through the end points of the major axis, and may be secant, tangent or exterior one to each other depending on the relative sizes of their radii and the length of the major axis. If they are secant (fig. 25 of the original text) and the centres of the large arcs are located on the points of intersection of the small ones, then this corresponds to the case of the second oval of Fray Lorenzo de San Nicolás [1639: Chapter LXXVIII], but now complying with the conditions of tangency. If they are tangent (fig. 26 of the original text) and the centres are aligned to form a 60° angle to the horizontal line, as Brizguz proposes, then the oval is the one by Vandelvira [c. 1580: fol. 18r]. If the two equal arcs are exterior to one other (fig. 27 of the original text) the centre of the large arcs can be placed anywhere, as long as they respect the tangency conditions. Nevertheless the drawing by Brizguz may cause confusion, as it seems that the oval passes through those centres (E and its symmetric) and that is not the case. The last layout proposed by the author (fig. 28 of the original text), in spite of its intention of "describing an oval given the major and minor diameters", copies the proposal of De La Rue, which does not follow an exact geometrical construction (fig. 20).

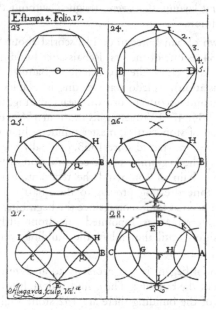

Fig. 20. Atanasio Genaro Brizguz y Bru, *Escuela de Arquitectura Civil* [1738: 17]

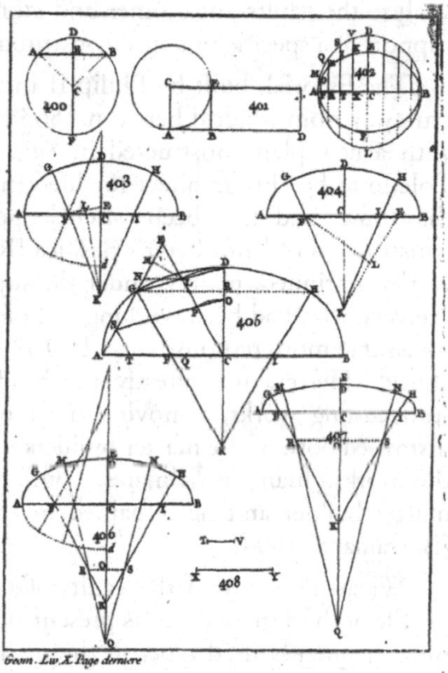

Fig. 21. M. Camus, *Cours de Mathématiques,* [1750]

The contribution by M. Camus [1750: vol. 2, 504-538] is a complete study of oval layouts for any given proportion for arcs of three and five centres, comprising thirty pages of the book [Huerta 2007: 234] (fig. 21).

As we have seen, the solution to the geometrical problem of laying out an oval fitting any given proportion would only be achieved at the beginning of the eighteenth century, in the *Compendio Mathematico* of Spanish author Tosca. We can hypothesize that perhaps the problem was not solved earlier because it was not necessary: builders had been constructing ovals by the trial-and-error method of adjustment. An analysis of the vaults at El Escorial will lend support to this argument, but it will also raise other doubts.

## Ovals for any given proportion in the Escorial

The study of the stone vaults at the Escorial shown in the present paper is part of the Ph.D. dissertation of the author [López Mozo 2009]. The work proved that one of the most renowned vaults in the building, the famous flat vault in the forechurch, is not the only outstanding contribution out of the problems tackled during its construction, and is perhaps not even the most interesting. A whole body of knowledge was developed: oval arches of all kinds were constructed; new solutions for pointed lunette and sail vaults were put forward, and, in particular, extradosed stone domes on drums, unprecedented in Spain. The general context of all these contributions is the aim of translating the Renaissance repertoire of vaults, which were mainly masonry brick built in Italy, to the language of stonecutting, carried out in the sixteenth century in Spain and France. This effort is reflected in the texts by De l'Orme [1567] and Vandelvira [c. 1580]. Moreover, original documents from the seventeenth century prove that a short treatise – now lost – was written at the Escorial worksite [Marías 1991], something easy to consider if we analyze the vaults and domes and stonecutting layouts that were designed without the support of a specific practical constructive tradition.

The Escorial, built by Philip II in the town of San Lorenzo de El Escorial about 50 km away from Madrid between 1563 and 1584, is one of the first Renaissance buildings with a new plan constructed in Spain. The king chose the Spanish Juan Bautista de Toledo to be "his architect for life" (my translation). Toledo was working in Naples at that time, and had been Michelangelo's second architect for two years during the construction of Saint Peter's Basilica [Rivera 1984: 47, 125]. After his death in 1567, the works continued, mainly under the supervision of stonecutting master builders. Juan de Herrera, who had been working at Toledo's office in Madrid since 1563, gradually began to assume more responsibility. In 1576, during the beginning of the construction of the church, Herrera was already the head of the chain of command: he reorganized the stonecutting works, removing most of the stone dressing process to the quarry, and dismissed both of the master builders who had been at the Escorial since the beginning of the works. Juan de Minjares, confidant of Herrera, was appointed sole stonecutting master builder and he remained in that position until the end of the building works [Bustamante 1994].

When the study of the vaults of the Escorial began, it was necessary to analyze the profile of the surbased vaults present throughout the building. The cylindrical vault is the most frequently used type: in the main level of the building there are 132 stone vaults; 106 of them are cylindrical and 69 surbased (fig. 22). All the significant vaults have been measured with a laser station to provide accurate data.

The existing groin and cloister vaults in the building prove that our masters knew how to reduce the height of an arch without changing the span, in order to correctly solve the intersection of the barrel vaults. We cannot determine if they used the layout by Dürer [1525], Serlio [1545], De l'Orme [1561] or Vandelvira [c. 1580], but the one by Serlio can be found in an original drawing – perhaps by Toledo – kept in the Library of the Royal Palace in Madrid [Bustamante et. al. 2001]. The method to elongate a semicircular arc was also known, based on the procedures just mentioned, but these were not specifically explained until the manuscript of the Spaniard Ginés Martínez de Aranda [c. 1600]. This case is actually built at El Escorial in the intersection of two small oblique barrel vaults with different, pre-fixed widths, behind the church cornice [Calvo 2002: 419-435].

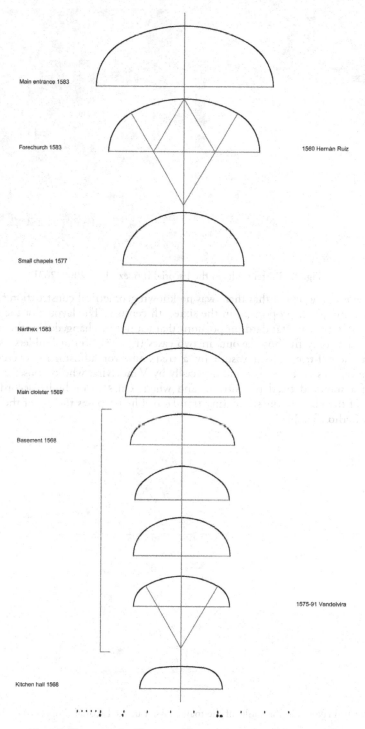

Main entrance 1583

Forechurch 1583

1560 Hernán Ruiz

Small chapels 1577

Narthex 1583

Main cloister 1569

Basement 1568

1575-91 Vandelvira

Kitchen hall 1568

Fig. 23. The most significant oval layouts in vaults of the Escorial,
drawn at the same scale [López Mozo 2009: 509]

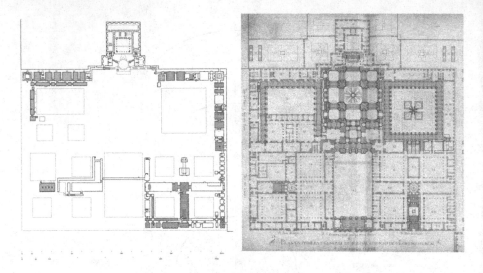

Fig. 22. Barrel vaults in the Escorial [López Mozo 2009: 263]

It is generally accepted that there was no known geometrical construction for drawing an oval for any given proportion in the sixteenth century. The layouts of the treatises at that time only present standard proportions that cannot be changed. But the oval figures at the Escorial only fit those layouts in two cases (fig. 23), so our builders were able to construct them fitting any measure by a trial-and-error adjustment, overcoming the limitations expressed by Serlio, and especially by Vandelvira, who proposes a single oval layout of a standard fixed proportion, and when it has to be built "according to the demands of the place in question" (my translation) he proposes the use of the "elliptical" layout by Serlio (fig. 24).

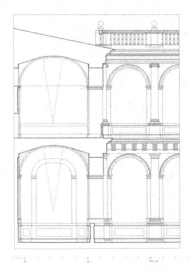

Fig. 24. Oval layouts in the vaults of the main cloister in the Escorial [López Mozo 2009: 283]

The analysis of the oval layouts of the vaults of the church would give unexpected conclusions. The surbased profile of the narthex vault fully fits an elliptical layout (fig. 25). However, for the sake of simplicity of the construction process alone, it would be more sensible to discard this option: on one hand, the mouldings of constant width in the arches and in the front walls cannot be laid out with ellipses; on the other hand, all the templates needed to dress the voussoirs of half the arch would be different; and, finally it is not possible to divide an ellipse into equal parts in order to distribute the voussoirs. These disadvantages of the elliptical layout versus the oval one would explain the fact that we only find built elliptical forms at the Escorial in small elements such as niches, lantern pilasters profiles (main dome, Fountain of the Evangelists) and in places where they were unavoidable (oblique barrel vaults behind the cornice of the church and surface intersections in groin and cloister vaults) [López Mozo 2009: 273-278].

Fig. 25. The Escorial, vault in the narthex

Once an elliptical layout for the narthex vault was discarded, the comparison of the measured points to the oval layouts of the treatises of the sixteenth century proved to be unsuccessful. Applying the modern method, described by Tosca in 1712, the oval layout that best fitted the vault cross-section showed an alignment of the centres forming a 45° angle to the horizontal line, a condition also satisfied in the determination of oval layouts in other church vaults (fig. 26). The recurrence of this situation made it necessary to consider the possibility that an oval layout for any given proportion with alignment of centres forming a 45° angle could have been known at the Escorial.

A simple mathematical deduction, set out in fig. 27, gave an easy graphical translation. The process was as follows: given an oval of semi-axes $a$ and $b$, it is possible to find the value $h$ which determines the distance of the centres to the vertical axis and to the horizontal springing line, resulting in an oval that passes through the final points of the axes.

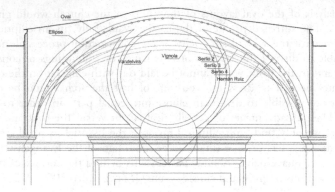

Fig. 26. Cross-section of the vault in the narthex,
comparing measured points and both elliptical and oval layouts

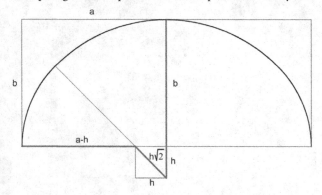

Fig. 27. Oval layout for any proportion with centres forming an angle of 45° to the vertical axis:
setting out [López Mozo 2009: 288]

$$a - h + h\sqrt{2} = h + b$$

$$a - b = 2h - h\sqrt{2}$$

$$a - b = h(2 - \sqrt{2})$$

$$h = \frac{(a-b)}{2-\sqrt{2}}$$

$$h = (a-b)\frac{2+\sqrt{2}}{(2-\sqrt{2})(2+\sqrt{2})} = (a-b)\frac{2+\sqrt{2}}{2} = (a-b)\left(1+\frac{1}{\sqrt{2}}\right)$$

$$h = (a-b) + \frac{(a-b)}{\sqrt{2}}.$$

Fig. 28 shows a graphical translation of the terms of the last mathematical expression. The corresponding process to lay out the oval is described by three steps in fig. 29.

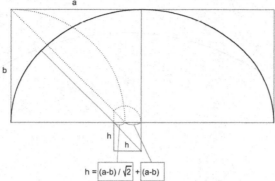

$$h = \boxed{(a-b)/\sqrt{2}} + \boxed{(a-b)}$$

Fig. 28. Oval for any given proportion with centres aligned forming a 45° angle to the horizontal line. Graphical translation

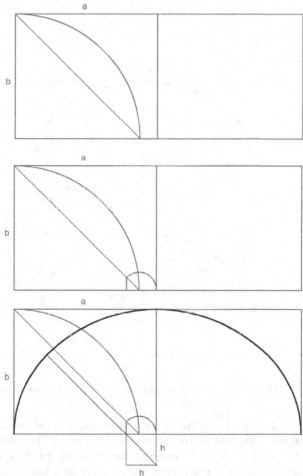

Fig. 29. Oval for any given proportion with centres aligned forming a 45° angle to the horizontal line. Layout process

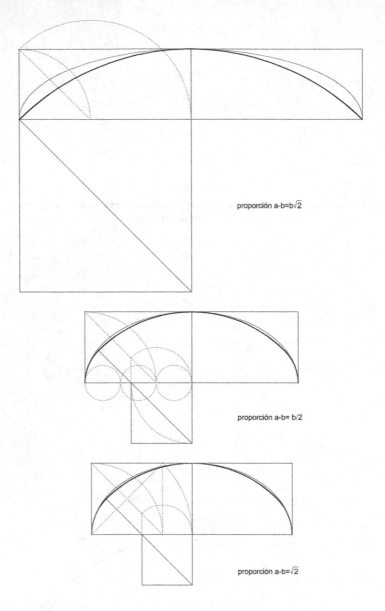

proporción a-b=b√2

proporción a-b= b/2

proporción a-b=√2

Fig. 30. Oval for any given proportion with centres at 45°. Specific cases.
The elliptical layout is also drawn

The layout just explained contains certain restrictions: it is not practical in cases with low proportions, since the first centre exceeds the segment for the span of the oval; the condition only holds for $a - b < b\sqrt{2}$, in which case the starting arch is at the extreme, its radius is zero, and the result is a segmental arch. In very low, but possible, proportions, the layout differs greatly from an ellipse. It would be more logical to make the arch starting from the proportion: $a - b = b/2$ (fig. 30).

Fernández Gómez [1994: 351, 378] describes this oval when listing current layout methods, without pointing out the condition satisfied by the centres, aligned at a 45° angle with respect to the axes. Neither the graphical drawing process nor the written explanation coincide with that offered in the present paper (fig. 31).

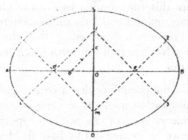

Fig. 31. Margarita Fernández Gómez [1994: 378]

## Conclusion

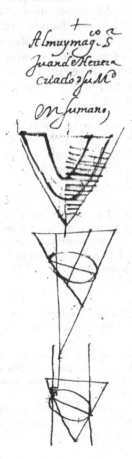

Judging by the number of times Serlio's oval layouts appear in subsequent treatises examined in this study, his enormously influential treatise enjoyed a very wide-spread dissemination. We find his proposals in the works of Hernán Ruiz [c. 1560], Jousse [1627], Fray Lorenzo de San Nicolás [1639], Simón García [1681], Derand [1643], Caramuel [1678], Milliet Dechales [1674] and Tosca [1712].

Vandelvira poses the problem of oval layouts in a manner similar to that of Serlio, but with a single possible oval, of fixed proportions; in the cases where the oval does not fit the given place, he proposes the elliptical layout posed by the Italian architect. It is strange that Vandelvira did not include the other three ovals by Serlio, especially since one of them has a √2 proportion and could fit the intersections of the cylinders forming the groin vaults. On the other hand, it is surprising that Vandelvira, who possessed abundant experience from his father's works, did not mention the possibility of drawing oval arches by approximating the centres. The work carried out in the construction of the vaults at El Escorial does offer evidence in this direction, since the majority of the ovals that were used do not fit either the fixed proportions mentioned in the treatises, nor an elliptical shape: the practical mastery evident in their execution – the skilful adjustment of the oval arches to the measures imposed by the worksite – seems to be present throughout the entire building.

Fig. 32. Drawings of conical sections attributed to Juan de Herrera (Archivo General de Simancas)

The precise geometrical construction necessary for drawing an oval adjusted to a given set of dimensions of the axes probably did not appear until 1712, in the *Compendio Mathematico* by Spanish author Tomás Vicente Tosca. It is the oval layout most frequently used nowadays. Two important later French authors, De La Rue [1728] and Frézier [1737], ignore this contribution by Tosca. The study of this topic with regards to his sources, dissemination and later influence has yet to be carried out. The oval layout put forth by Frézier is less interesting than the one by Tosca, since it is a specific case. However, it solves the problem for a given inclination of 60° of the line segment that joins the centres.

This work puts forward the possibility that a correct geometrical method for laying out ovals for any given proportion might have been known at the Escorial at around 1576. Nevertheless, no evidence exists of this fact; it is hypothesized based on observation of the second stage of the works, carried out by Juan de Herrera and his master builder Juan de Minjares. If this hypothesis is correct, it seems strange that the method would not have spread. However, it might be that the lack of a subsequent echo was due to the fact that builders did not need such a precise tool: they laid out oval arches for any given place by means of trial and error. On the other hand, Herrera's solid mathematical and graphical knowledge lends support to this hypothesis. Herrera, who founded the Academy of Mathematics in Madrid in 1582 at the behest of Philip II, possessed a library containing dozens of books related to mathematics, including several editions of works by Euclid, Archimedes and Apollonius [Aramburu-Zabala 2003]. A set of drawings containing conical sections found within a document sent to Herrera by the Secretary of Philip II are attributed to Herrera himself [García Tapia 1990] (fig. 32).

A new graphical description of a layout that is not widely known today is shown in this paper: it is the one that draws an oval beginning from its axes, satisfying the condition that the line segment that connects the two centres form a 45° angle with the minor axis. This is a particular case, like that by Frézier, but for a different angle.

*Acknowledgments*

This paper is part of the research project "Ashlar construction in the Mediterranean basin and Atlantic area. Analysis of built heritage", which is funded by the Ministry of Science and Innovation of Spain (BIA2009-14350-C02-01). I wish to express my gratitude to Enrique Rabasa for his kind permission to use his work on Gelabert's manuscript and to José Calvo Lopez for making available an electronic facsimile of the 1515 edition of Serlio's Books I and II.

*Notes*

1. Vandelvira proposes drawing the "elliptical" line of the *arco painel* by taking points in sets of three and drawing the circular arc that passes through them [c. 1580: Section 22; Barbé-Coquelin de Lisle 1977: Vol. I, 56].
2. The term "carpanel" might have originated in the French *anse de panier*, which refers to a basket handle, in the same manner as the term *ansapaner* used by the Majorcan Gelabert [Rabasa 2011]. Juan Bautista de Toledo, the first architect for the Escorial, used the term *ançarpanel* in 1565 [Portabales 1945: XXXIV-XXXVII; López Mozo 2009].
3. According to Antonio Bonet Correa, the first four chapters are either a word-for-word copy or at least very faithful extracts of the text and ideas found in the manuscript by Rodrigo Gil de Hontañón. Other authors attribute to him chapters 1 through 6 at the most [1991: 14]. The topic on ovals is found in Chapter 20.
4. Margarita Fernández Gómez describes the layout, indicating that it starts out from the axes of the oval [1994: 351].

# References

ARAMBURU-ZABALA HIGUERA, Miguel Ángel, Ana CAGIGAS ABERASTURI and Celestina LOSADA VAREA. 2003. *Biografía de Juan de Herrera.* Santander: Fundación Obra Pía Juan de Herrera.

BACHOT, Ambroise. 1598. *Le Gouvernail d'Ambroise Bachot capitaine ingenieur du Roy, Lequel conduiraie curieux de Geometrie en perspective dedans l'architecture des fortifications, machines de guerre et plusieurs autres particularites et contenues.* Melun: Chez l'Auteur. [Electronic facsimile in Gallica, http://gallica.bnf.fr/].

BARBE-COQUELIN DE LISLE, Geneviève. 1977. *El tratado de arquitectura de Alonso de Vandelvira.* Albacete: Caja de Ahorros Provincial de Albacete.

BONET CORREA, Antonio. 1991. Simón García, tratadista de arquitectura. Pp. 13-18 in *Compendio de Architectura y simetria....por Simon Garcia.* Valladolid: Colegio Oficial de Arquitectos de Valladolid.

BRIZGUZ Y BRU, Atanasio Genaro. [1738] 1804. *Escuela de Arquitectura Civil.* Valencia: Joseph de Orga.

BUSTAMANTE GARCÍA, Agustín. 1994: *La octava maravilla del mundo (Estudio histórico sobre el Escorial de Felipe II).* Madrid: Alpuerto.

―――. 2001. (et. al.) *Las Trazas de Juan de Herrera y sus seguidores.* Catalogue of the exhibition organized by Patrimonio Nacional and Fundación Marcelino Botín.

CALVO LÓPEZ, José. 2002. La semielipse peraltada. Arquitectura, geometría y mecánica en las últimas décadas del siglo XVI. Pp. 419-435 in *Proceedings of the Symposium El Monasterio de El Escorial y la arquitectura.* Madrid: Ediciones Escurialenses.

CAMUS, M. 1750. *Elémens de géométrie théorique et pratique (Cours de mathématique, Seconde Partie).* Paris: Durand [Electronic facsimile in *Gallica*, http://gallica.bnf.fr/].

CARAMUEL DE LOBKOWITZ, Juan. 1678. *Arquitectura civil recta y oblicua...* Vigevano: Imprenta Obispal [Electronic facsimile in Internet Archive, http://www.archive.org/details/texts].

DE LA RUE, Jean Baptiste de la. [1728] 1764. *Traité de la coupe des pierres.* Paris: Chez Charles-Antoine Jombert.

DE L'ORME, Philibert. 1567. *Le premier tome de l'Architecture.* Paris: Federic Morel.

DERAND, François. 1643. *L'Architecture des voûtes.* Paris: Chez Sebastien Cramoisy. [Electronic facsimile in Architectura, http://architectura.cesr.univ-tours.fr/index.asp].

DÜRER, Albrecht. 1525. *Underweysung der messung mit dem zirkel und richtscheyt ....* Nürnberg. [Electronic facsimile of the 1538 ed. in *Bibliotheca Perspectivae*, www.imss.fi.it].

FERNÁNDEZ GÓMEZ, Margarita. 1994. Trazas de óvalos y elipses en los Tratados de Arquitectura de los siglos XVI y XVII. Pp. 335-378 in *La formación cultural arquitectónica en la enseñanza del dibujo.* Actas del Quinto Congreso Internacional de Expresión Gráfica Arquitectónica, Las Palmas de Gran Canaria 5-7 May 1994. Universidad de las Palmas de Gran Canaria, Departamento de Expresión Gráfica y Proyectación Arquitectónica.

―――. 2000. *Estudio de los Tratados XIV y XV del Compendio Matemático del Padre Tosca.* Valencia: Universidad Politécnica de Valencia.

FRÉZIER, Amédée-François. 1737-39. *La théorie et la pratique de la coupe de pierres et des bois pour la construction des voûtes et autres parties des bâtiments civils et militaires, ou traité de stéréotomie à l'usage de l'architecture.* Strasbourg/Paris: Charles-Antoine Jombert.

GARCÍA TAPIA, Nicolás. 1990. *Ingeniería y arquitectura en el Renacimiento español.* Valladolid: Universidad de Valladolid.

GARCÍA, Simón. 1990. *Compendio de arquitectura y simetría de los templos...* Valladolid: Colegio Oficial de arquitectos de Valladolid (Facsimile ed. of a 1681 ms).

GELABERT, Joseph. 1977. *De l'art de picapedrer.* Palma de Mallorca: Instituto de Estudios Baleáricos. (Facsimile ed. of a 1653 ms. in the library of the Consell Insular de Mallorca, Palma de Mallorca).

GENTIL BALDRICH, José María. 1996. La traza oval y la sala capitular de la catedral de Sevilla. Una aproximación geométrica. Pp. 77-147 in *Quatro edificios sevillanos.* J. A. Ruiz de la Rosa et al., eds. Seville: Colegio Oficial de Arquitectos de Andalucía, Demarc. Occidental.

HUERTA, Santiago. 2007. Oval Domes: History, Geometry and Mechanics. *Nexus Network Journal* 9, 2: 211-248.

JOUSSE, Mathurin. 1642. *Le secret d'Architecture*. La Flèche: Georges Griveau. [Electronic facsimile in Architectura, http://architectura.cesr.univ-tours.fr/index.asp].

———. 1702. *L'art de Charpenterie . . . corrigé et augmenté . . . par M. de La Hire. 1627.* Paris: Thomas Moette [Electronic facsimile in Gallica, http://gallica.bnf.fr/].

LÓPEZ MOZO, Ana. 2000. Las bóvedas de los sótanos de Poniente del Monasterio de El Escorial. Pp. 615-621 in *Actas del III Congreso Nacional de Historia de la Construcción*. Madrid: Instituto Juan de Herrera.

———. 2003. Planar vaults in the Monastery of El Escorial. Pp. 1.327-1.334 in *Proceedings of the First International Congress on Construction History*. Santiago Huerta Fernández, ed. Madrid: Instituto Juan de Herrera.

———. 2005. Bóvedas cilíndricas en el Monasterio de El Escorial: dos ejemplos de lunetos. *Actas del Cuarto Congreso Nacional de Historia de la Construcción*. Santiago Huerta Fernández, ed. Madrid: Instituto Juan de Herrera.

———. 2009. Bóvedas de piedra del Monasterio de El Escorial. Ph.D. dissertation, Universidad Politécnica de Madrid.

MARÍAS FRANCO, Fernando. 1991. Piedra y ladrillo en la arquitectura española del siglo XVI. Pp. 71-83 in *Les chantiers de la Renaissance*. Actes des colloques tenus à Tours en 1983-1984. Jean Guillaume, ed. París: Picard.

MARTÍN TALAVERANO, Rafael. 2007. La bóveda del sotocoro del convento de Santo Tomás (Ávila). Pp. 649-658 in *Actas del Quinto Congreso Nacional de Historia de la Construcción*. Santiago Huerta Fernández, ed. Madrid: Instituto Juan de Herrera.

———. 2009. Two Flat Ribbed Vaults in San Juan de los Reyes (Toledo, Spain). Pp. 31-38 in *Proceedings of the Third International Congress on Construction History*. Cottbus: Chair of Construction History and Structural Preservation of the Brandenburg University of Technology Cottbus.

MARTINEZ DE ARANDA, Ginés. 1600 c. *Cerramientos y trazas de montea*. (Ms. in the library of the Servicio Histórico del Ejército, Madrid. Facsimile ed., Madrid, Servicio Histórico del Ejército-CEHOPU, 1986.)

MILLIET DECHALES, P. Claude François. 1674. *Cursus seu mundus mathematicus*, 2 vols. Lyon: Anissonm.

NAVASCUÉS PALACIO, Pedro. 1974. *El libro de arquitectura de Hernán Ruiz el Joven. Estudio y edición crítica*. Madrid: Escuela Técnica Superior de Arquitectura.

PALACIOS GONZALO, José Carlos. 2003. Trazas y cortes de cantería en el Renacimiento español. Madrid: Munilla-Lería. (1st ed., Madrid: Ministerio de Cultura, 1990).

———. 2003. Spanish ribbed vaults in the 15th and 16th centuries. Pp. 1547-58 in *Proceedings of the First International Congress on Construction History*. S. Huerta, ed. Madrid: Instituto Juan de Herrera.

PORTABALES PICHEL, Amancio. 1945. *Los verdaderos artífices de El Escorial y el estilo indebidamente llamado herreriano*. Madrid: Gráfica Literaria.

RABASA DÍAZ, Enrique. 2000. *Forma y construcción en piedra. De la cantería medieval a la esteorotomía del siglo XIX*. Madrid: Akal.

———, ed. 2011. *El manuscrito de cantería de Joseph Gelabert*, Madrid: Fundación Juanelo Turriano.

RIVERA BLANCO, Javier. 1984. *Juan Bautista de Toledo y Felipe II. La implantación del clasicismo en España*. Valladolid: Secretariado de Publicaciones de la Universidad de Valladolid.

ROJAS, Cristóbal de. 1598. *Tres tratados sobre fortificación y milicia*. Madrid: Luis Sanchez.

ROSSELLÓ, Vicenç M. 2004. Tomàs V. Tosca y su entorno ilustrado en Valencia. Obra autógrafa y atribuciones. *Ería* **64-65**: 159-176

RUIZ, Hernán el Joven. c. 1560. *Libro de arquitectura*. Ms. R.16, Biblioteca de la Escuela Técnica Superior de Arquitectura, Madrid.

SAN NICOLÁS, Fray Lorenzo de. 1989. *Arte y uso de architectura. Primera parte* (1639). Rpt. Valencia: Albatros Ediciones.

SERLIO, Sebastiano. 1545. *Il Primo libro d'Architettura di Sebastiano Serlio. . .* Paris: 1545.

TOSCA, Tomas Vicente. 1712. *Compendio matemático...* Valencia: Antonio Bordazar [Electronic facsimile of the 1727 ed. in Sociedad Española de Historia de la Construcción, http://gilbert.aq.upm.es/sedhc/index.htm].

VANDELVIRA, Alonso de. c. 1580. *Exposición y declaración sobre el tratado de cortes de fábricas que escribió Alonso de Valdeelvira por el excelente e insigne architecto y maestro de architectura don Bartolomé de Sombigo y Salcedo, maestro mayor de la Santa Iglesia de Toledo*. Ms. R.10, Biblioteca de la Escuela Técnica Superior de Arquitectura, Madrid.

## About the author

Ana López Mozo is an architect. Her Ph.D. dissertation was awarded the Extraordinary Doctoral Prize of the Polytechnic University of Madrid in 2011. She is Professor of Geometry, Architectural Drawing and Construction History at the School of Architecture of the Polytechnic University of Madrid. She has also lectured in graduate courses on architectural restoration and computer graphics in architecture at Madrid and at San Pablo-CEU University. Her research focuses on stereotomy, arches, vaults and domes and has appeared in international conference proceedings, journals and books such as *Cantería renacentista en la catedral de Murcia* (Murcia: Colegio Oficial de Arquitectos de Murcia, 2005).

Giuseppe Fallacara*
*Corresponding author
Facoltà di Architettura
Politecnico di Bari
via Orabona, 4
70125 Bari - Italia
gfallacara@hotmail.com

Fiore Resta

Facoltà di Architettura
Politecnico di Bari
via Orabona, 4
70125 Bari - Italia
fiore.resta@email.it

Nicoletta Spallucci

Facoltà di Architettura
Politecnico di Bari
via Orabona, 4
70125 Bari - Italia
nicla.spa@email.it

Luc Tamboréro

Ecole Nationale Supérieure
d'Architecture Paris-Malaquais
14, rue Bonaparte
75272 Paris - France
mecastone@hotmail.fr

Research

# The Vault of the Hôtel de Ville in Arles

**Abstract.** The vault of the Arles City Hall, or Hôtel de Ville, represents the architect's revenge on the corporations of masons. Completed in 1676, the relationship of span to rise of the vault make it the boldest work of masonry in Europe, and while this complex vault appears to be a unitary structure, two vaults actually share the work, leaning against each other on the big arch. Bibliographic and archival research showed that surveys of the vault were missing; the only ancient survey was lost in the 1970s. A recent survey campaign made an analysis possible, leading to a hypothesis about the architect Mansart's choices, based on hypothetical solutions to resolve the formal construction issues of the vault. A connoisseur of geometry and optics, Mansart knew that the human eye was unable to perceive the exact geometry of a surface. Knowing that he couldn't control the intersections of vault portions and then the joints of the rows in the space, he introduced a solution that involved drawing the intersections in plan and then projecting them on the vault to obtain the spatial intersections. No further constructions of the bold type followed the vaulted space in the Hôtel de ville, because the spatial research that linked the new discipline (stereotomy) to the quality of architectural space had by then come to an end.

Keywords: masonry construction, vaults, stereotomy,

*La voûte de l'Hôtel de ville d'Arles est le chef d'oeuvre de la stéréotomie française*
[Perouse de Montclos 1983: 123-126]

## Introduction

The vault of the *Hôtel de ville*, or city hall, in Arles symbolises the revenge of the figure of the architect on the guilds of masons. Completed in 1676, the relationship between the span and the height of the vault of the vestibule make it still today the most daring creation in dressed stone in all Europe. It is ingeniously conceived to give the illusion of being a single structure, while in reality it is composed of two vaults that share the work, leaning against each other on the large arch. The special feature of the stereotomy consists in the fact that the surface was not conceived as an assembly of portions of canonic geometric volumes, but is an expression of a process of laying out lines in plan (a wire frame) and projecting this onto the vault.

It was just in this period, in December 1671, that the Académie Royale d'Architecture was founded, and thus the vault was built at a moment in history when the figure of the architect clashed with that of the *tailleur de pierre*, or stonecutter. This conflict can be dated back to 1644, to the debate between Curabelle, who represented the guild of masons, and the geometer Girard Desargues. The vault represents the last work of stereotomy of this level of complexity ever constructed. The separation of roles introduced by the Académie led architects and engineers to prefer forms that were sobre and regular, for aesthetic reasons and above all, for structural reasons.

## The history of the Arles Hôtel de ville

The *Hôtel de ville* was built to satisfy the need for a space of sufficient size to house the Arles city council meetings. Since the Middle Ages, the city administration in Arles had been under the control of four consuls – two noble and two bourgeois chosen from among the members of the city council – who met annually. The life of the city was regulated by the resolutions of this council. With the organisation of the city administration at the end of the fourteenth century, the city council realised that they needed a place of their own in which to meet.

Construction on the new city hall (fig. 1) lasted almost twenty years, with the building entrusted to the greatest architects of the day, including La Valfenière, Puget, Jacques Peytret and Jules Hardouin-Mansart. Following demolition of the old city hall, some weeks after construction had begun work was stopped for a change in the design brought about by the decision to purchase the house belonging to Alphante di Bibion, located to the east of the site of the city hall, where the clock tower was. This modification led to an increase in both width and height of planned building.

Fig. 1. The present day façade of the Arles *Hôtel de ville* on the Place de la République

The design presented by La Valfenière was fairly different from that later adopted. The poor quality of the work made it necessary to demolish what had previously been constructed all the way down to the foundations.

Only Mansart dared to incorporate the old clock tower in its entirety within the perimeter of the new building. The tower was built in the form of an irregular trapezoid placed slightly obliquely with respect to the alignment of the two spaces. Mansart's design called for a large central space conceived without intermediate supporting piers, with a vault that covered it entirely with a height of only 1 *canne*, a local unit of length equal to just over 2 m. At that time, architects and stonecutters had very different views about and training in geometry. Mansart chose Jacques Peytret, a painter, engraver and architect, to accompany him to Bèziers, explaining to him along the way the design and the techniques to be used in its execution.

When he returned to Arles, Peytret was already carrying the drawings of the *panneaux* for the construction of the vault. Unfortunately, he did not entirely follow the architect's design, believing it to be too daring: he decided to raise the vault by one and a half *pan* (about 30 cm).

On 29 August 1675 the last stone was placed in the vault, and the centering removed.

## The debates at the Académie Royale d'Architecture

The *Académie Royale d'Architecture* was created in Paris in 1671, thanks to the efforts of Jean-Baptiste Colbert and the architect François Blondel. The Académie debates concerned the issues of the day: the current status of the architectural profession, and the problem of structural stability of vaults. The vault in Arles was the focus of debates regarding the relationships between mechanics, geometry, and architecture. Its irregularity made it impossible to compare to other works of stereotomy.

One key figure in the history of stereotomy and mechanics applied to construction is Philippe de La Hire, professor at both the *Académie Royale d'Architecture* and the *Collége Royale*, a member of the Paris *Académie Royale des Sciences*, and authoritative advisor to Louis XIV. De La Hire addressed the problem of vaults both by examining the aspects relative to geometry and stonecutting, and by studying the problem of determining the thrust.

Consider that the breaking point of the arch is located close to that angle beyond which friction is no longer able to guarantee equilibrium; this is found at approximately 45°. The stones of the upper part, comprised between the two joints of rupture behave as a single block, and the kinematics is like that of a large keystone that pushes against the sides of the vault.

The problem that provoked the discussions could have been one of those given to the students of the *Académie Royale d'Architecture:* design a large vaulted vestibule, comprising two adjacent bays that together form a space some 16 m square, without intermediate supports, with a rise equal to a ninth of the span.

## Analysis of the vault of the Hôtel de ville

The vaulted space of the Arles city hall aroused quite a bit of interest on the part of the professors in the Faculty of Architecture at the Politecnico di Bari, who decided to use it as a case study in a degree program. Following a survey campaign, it was possible to formulate several hypotheses for an analysis that might make it possible to retrace and understand Mansart's design principles. The hypotheses are obviously subjective, and are dictated by the deductions and interpretations deriving from the survey.

The interpretation of these kinds of proportional systems, which are as legible in plan as they are in elevation, go from the decomposition of the work in *cannes* and *pans* to the recomposition of the parts according to the rules of the golden rectangle, finally arriving at the intuition of the purely sensory equilibrium by means of perspective corrections.

The unit of measurement used by classical architects to proportion the whole is the module, which is determined by the diameter of the column shaft measured at its base: this is the fundamental building block of the entire system. Once that unit of measure (the module) has been established, all architectural elements, frames, mouldings, and so forth are determined as whole multiples or simple fractions of the module. The dimension established is used as a reference for all others obtained according to pre-determined ratios.

On 14 November 1583, by order of Henri III, it was decreed that all the various weights and measures of Provence were to be unified in conformance with the other units of measure of southern France. The unit of measure became the *canne*, equal to 2.044 m. Each *canne* was divided into 8 *pans*, each equal to approximately 0.25 m.

## Analysis in modules

It has emerged that in plan the proportioning module measures exactly equal to 8 *cannes* by 8 *cannes*.

Analysis of the proportions in the elevations has confirmed what was indicated in archival documents: the facade in *Plan de la Cour* is larger than that in *Place de la République*, measuring 15 *cannes*, 4 *pans* as opposed to the 14 *cannes*, 6 *pans* of the present-day main façade (fig. 2).

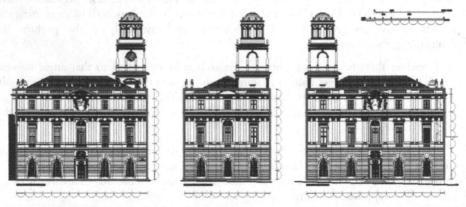

Fig. 2. Analysis of the facades in Place de la République, in rue de l'Hôtel de Ville and in Plan de la Cour, in *canne* and *pan*, the ancient French units of measurement

Two golden rectangles can be identified in the structure of the vault, one in which the large vault is inscribed, and the second for the small vault. The two rectangles follow two proportional systems: one based on sub-modules of the rectangle itself, the other on the system of the *cannes* (fig. 3). The two units of measure closely approximate each other, differing by only 2 *menu* (0.032 m). In particular, the rectangle of the large vault measures 8 x 5 modules, where each module is equal to the eighth part of the base of the rectangle. The sum of the heights of the largest rectangle and the rectangle that circumscribes the small vault is 8 modules. When read using the system of *cannes*, the largest rectangle measures 8 *cannes*, 2.5 *pans* by 5 *cannes*, 1 *pan*. The small rectangle measures 3 *cannes*, 1 *pan* (fig. 4).

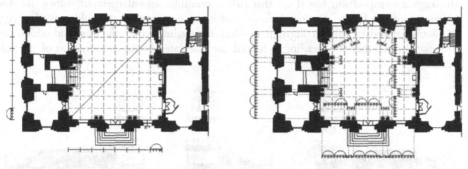

Fig. 3. Analysis of the plan in *canne* and *pan*

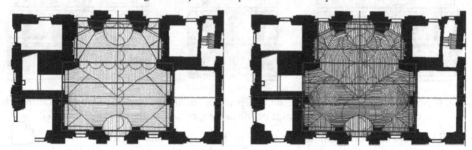

Fig. 4. Analysis of the vault with *canne* and *pan*

The system is in base 8: the interior columns are 2 *cannes* high, equal to 16 *pans*; the rise is equal to an eighth of the span of the vault's rise.

## Analysis in terms of the golden section

Beginning with the plan of the entire Hôtel de ville it can easily be seen that the whole composition is governed by a golden rectangle whose base is the exterior perimeter of the north facade on Plan de la Cour and whose height the exterior perimeter of the facade on the Rue de l'Hôtel de ville (fig. 5).

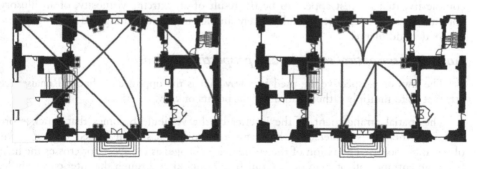

Fig. 5. Analysis of the plan using the golden rectangle

The golden rectangle is also recognisable in the present-day rear facade. An emblematic characteristic that has emerged from the examination of the elevation is the fact that the rectangle is not immediately legible on the facade on Place de la République: Although a composition based on this rule is credible, an enlargement shows that the perimeters of the facade do not perfectly match the geometry of the golden section (fig. 6). Doubts raised by this incongruence, which is puzzling to say the least and which can lead us to think that this reading is forced, are alleviated by an examination of the rear facade.

Fig. 6. Analysis of the facades in Place de la République, in rue de l'Hôtel de Ville and in Plan de la Cour, with the golden rectangle or "divine proportion" as described by Fibonacci

The elevation on Plan de la Cour in fact corresponds perfectly to the rigorous construction of the divine proportion. This observation takes on further significance thanks to archival research which has made it possible to understand that the design of the present-day rear facade was originally the main facade designed by the brother of Jules Hardouin-Mansart, Michael. This allows us to understand why the southern facade does not correspond exactly to the golden rectangle. Obviously it is not surprising to associate the elevation on Rue de l'Hôtel de ville with a perfect square. It should be recalled that the *ad quadratum* method of proportioning, with the square as the figure of reference, was used by architects from ancient times to determine building proportions.

The vault of the Arles Hôtel de ville, and more broadly, all of the vaulted vestibule, fully responds to the Baroque requirement that the entire work must resound with complexity, that it must appear to be the result of an extreme virtuosity, of an illusory appearance, at once exuberant and severely simple, aimed at creating and exalting a state of wonder and marvel.

## Analysis of the optical solutions and perspective illusions

The laws of perspective are used in a way that is the opposite of that of Renaissance art, that is, to multiply rather than unify the points of view.

The spatial arrangement of the interior of the vestibule becomes that much more amazing when it is understood how much attention has been placed on particular observation points. The vision of the entire space changes as the spectator crosses the hall: from the entrance off of Plan de la Cour (fig. 7), looking through the interior vestibule, the capitals placed in front appear exactly straight, in a view that is almost axionometric.

Fig. 7. View into the hall from the entrance of Plan de la Cour, where the column capitals are seen in an axionometric view

Fig. 8. Perspective view of the capitals of the inner south wall

As one gradually moves from the entrance off of Plan de la Cour towards the entrance off of Place de la Republique, crossing the vestibule longitudinally, the view becomes perspective (fig. 8), until one arrives under the great arch that subdivides the two vaults, where the view shows the capitals to be actually tilted (distorted view) (figs. 9, 10).

Fig. 9. View of the capitals in showing their actual distorted form

Fig. 10. Close-up of the capitals showing their actual distorted form

The changing way in which the whole composition is viewed as one moves through it, which only occurs when crossing the axis from north to south, confirms the hypothesis that the main facade was originally that located on the north side, in Plan de la Cour. This notion is also supported by the position and shape of the adjacent spaces: the doors of the rooms to the north and south have different dimensions, but on site, the eye is fooled into thinking that they are identical. The survey made of the interior elevations of the vestibule have made evident a notable difference between the doors close to the entrance off of Place de la République and those close to Plan de la Cour, of which both the frames and the stone jambs are actually higher and wider. This effectively constitutes an optical strategem so that when the spectator goes down the staircase (fig. 11) or stands under the bust of Mistral (fig. 12) he is presented with a perspective view and the doors appear exactly the same.

Fig. 11. Interior elevation looking towards the staircase. From this point there is no apparent difference between the doors

Fig. 12. Interior elevation facing the staircase. From this point there is no apparent difference between the doors

The aesthetic effect of the vault that Mansart was aiming for was that of making the vaulted space a *unicum*: all of the horizontals appear continuous and unbroken to the eye of the observer. This effect is heightened thanks to specific aesthetic devices, for example, covering the vertical joints with mortar and emphasising the horizontal joints. The horizontality is also obtained by means of some false grooves that do not follow the way the stone is cut: the eye does not see any kind of discontinuity.

Entering the vestibule from the present-day main entrance off of Place de la République, one has the impression that the hall is much smaller than its sixteen meters, and above all that it is shorter and dilated towards the sides (fig. 13). The flattened arch seems to have a very shallow rise, and to 'fall' heavily on those who cross the southern threshold. The optical effect is due to the continuous horizontality of the joints in the large vault, which gives a sense of 'pressing down'.

Fig. 13. Seen from the main entrance, the vault seems to have very shallow rises to "fall" heavily on those who pass through the south entrance. The optical effect is due to the continuous horizontal lines on the large vault, which create a strong sense of "crushing"

Entering instead from Plan de la Cour, the former main entrance, the vestibule appears to be dilated, the vault seems to be higher than it actually is, as though it were thrust upwards (fig. 14). This opposite optical effect is made possible by the continuous lines of the central part of the vault, which turn out to be, only in this point, perpendicular to the entrance, giving the idea of straight lines that mark the spectator's path and draw the eye towards infinity. This is an effect that Mansart's work shares with other coeval Baroque creations.

Looking with a critical eye at the portion of the vault that lies over the first steps of the staircase, the ruled surfaces seem to be exactly divided into two clearly distinct parts by the stone that begins at the vertex of the groin, which becomes the keystone, even though it is not in the middle (fig. 15). Looking along the straight line that passes through the vertex of the groin, which separates the two portions of the conoid, it appears that the perspective vanishing point of the line falls exactly on the middle axis of the wall at the end of the stairs, where the statue of Venus is located, while the two conoids could have two different vertices on the axis itself.

Fig. 14. Entering from Plan de la Cour, the former main entrance, the vestibule is dilated, and the vault seems to be higher than the actual measurement, as though soaring upwards

Fig. 15. The striped surface appears to be divided into two different parts from the keystone. Looking along the line passing through this voussoir, which separates the two portions of the conoid, it appears that the point of convergence of the straight lines falls exactly on the median axis of the wall at the bottom of the stairs, where is the statue of Venus is located, while the two conoids could have two different vertices, on the axis itself

While surveying the base of the statue, we noted a decentralisation of the sculpted point (fig. 16). This supports our geometric theory about the vanishing point, reflecting Mansart's Baroque leanings and sense of scenography. The point is decentralised because it is an intermediate point on the axis.

The reasoning begins with the collocation of the work in the Baroque period, when the upper floor was used as a ballroom. Thus one of the main points of observation of the vault was located precisely on the staircase.

Fig. 16. We noticed on the statue base a decentralization of the sculpted point: this corroborates our theory about the geometric perspective, in defense of the Mansart's Baroque scenocographic aesthetic. The point is decentralized because it's an intermediate point on the axis

The study of light is one of the fundamental characteristics of Baroque architecture. Light assumes an almost paramount importance during this period, and becomes a means for overcoming the static nature of architecture. The desire of the artist was that his work be seen differently according to the various hours of the day or night. Mansart's design goal was that, thanks to the high-quality *taille de pierre*, stonecutting, used to prepare the stones of the vault in Arles, at 12:00 a glancing light would enter the space, optimum for observing all the edges and spatial arcs of the vault and the material, a stone from Fontvieille, and that the faint glimmering would bring out the stone's true colours (fig. 17).

Fig. 17. View of the vault of the hall in the Hôtel de ville of Arles

Our analysis did not end with possible design interpretations, but went on to an examination of the geometry of the surface of the vault.

## The survey of the vault of the Hôtel de ville

The obvious difficulty of measuring the vault – characterised by convex and concave curves, and elements that project and rotate – using tradition survey methods, led to the decision to carry out the entire survey by means of a laser scanner. Other factors which led to that decision were:

– laser technology is essentially unaffected by the motion of passers-by and poor lighting;
– laster technology makes it possible to acquire a large number of points in a uniform manner to define the geometry of the structure without the need to predetermine the elements to be surveyed;
– the acquisition of a dense point cloud makes it possible to construct a three-dimensional model and take advantage of innovatiove systems of representation.

The digitalisation was performed by means of the measurements of the positions of a large number of points. The technique of laser scanning, also known as LIDAR (light detection and ranging), is based on the method of measuring the distances by means of electromagnetic waves. In the case of a survey with the laser scanner, there is no possibility of choosing the points to be surveyed, as in general it is only possible to define the area to be measured and the desired density of points. Once these parameters are defined, the acquistion of data is completely automatic. The result of the survey is a point cloud that is quite dense but with a random distribution of points on the object.

### *Editing the point cloud obtained by the laser scanner*

Once the data has been obtained by the laser scanner (the point cloud), it is necessary to proceed to the elaboration of the information obtained. The first step is the careful creation of a mesh that will guarantee the required level of precision and detail. The specific software for elaborating the data includes commands for regulating the maximum number of triangles or the minimum distance between adjacent points. However, these tools must be used with caution, because while on the one hand they help to eliminate data that is redundant, on the other there also exists the risk of eliminating information in places where redundant information is helpful. Once created, the mesh must be checked for holes, which can subsequently be minimised or eliminated.

It is possible to derive a series of sections directly from the point cloud obtained from the survey: in our case, twelve such sections were derived in the vertical plane, while fifty-two were derived in the horizontal plane. The sections were organised into small groups, and once detailed editing was performed on each, it was possible to import into a new file all the reworked, reformatted sections that had been chosen for the creation of the final model.

As mentioned, the point cloud had been sectioned by fifty-two horizontal planes and twelve vertical planes. The fifty-two horizontal planes were perpendicular to the vertical of the laser scanner, that is, orthogonal with respect to the $z$-axis of the global reference (fig. 18). This established the contour lines of the vault; these were useful for the subsequent analysis of the vault's deformations and geometry.

Fig. 18. Horizontal section obtained from the points cloud

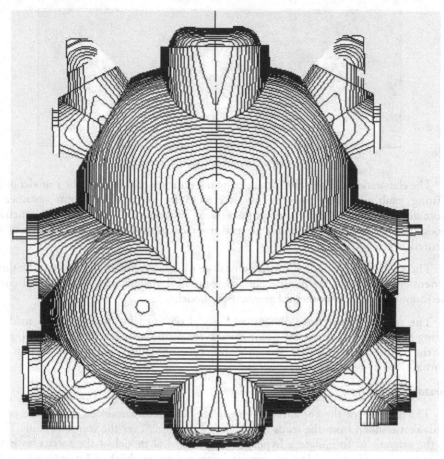

Fig. 19. Horizontal sections which shows the real surface of the vault

The contour lines made it possible to see that the vault showed a slight sag due to settlement of the cylindrical vault generated by the large arch. The directrixes of a cylindrical surface are parallel lines, whereas the directrixes of the started out parallel at the base of the vault and began to converge close to the apex (fig. 19). The subsequent flattening of the vault is in any case quite small (on the order of a tenth of a centimeter). This result confirmed the decision to carry out the survey by means of the laser scanner; no manual survey could possibly have made such a small deviation evident.

The twelve longitudinal sections were derived with planes parallel to the $x$-$y$- and $y$-$z$-axes of the global reference system. These planes allowed us to define the shapes of the vault's surface (fig. 20). At this point we deduced that the vault is actually composed of five different types of vaults: two barrel vaults and *anse de panier*, or basket handle shaped, lunettes. The main barrel vault is terminated at each end with two hemispherical vaults (*cul de four*) and is intersected by a second, transversal barrel vault, which in its turn terminates in a large lunette. We thus have four lunettes at the center of each side and four double lunettes in the corners.

Fig. 20. Vertical section obtained from the points cloud

The elaboration of the point cloud led to the creation of three models: a model of the existing vault obtained from the mesh created from the cloud point obtained; a geometrical model that approximates the existing situation; and a theoretical geometrical model that unites the information gleaned from documentary sources with that of the theoretical geometrical model.

The methodology adopted consisted of three phases: careful examination of the three-dimensional model of the point cloud; a geometrical analysis of each individual curve; the formulation of a hypothetical geometrical model.

The analysis of the three-dimensional model obtained by the survey was made by taking horizontal sections of the point cloud every 5 cm starting at the (horizontal) plane of the vault's impost. The sequence of points obtained with each section was reinterpreted by vectorization.

## Geometric analysis

The drawing of the contour lines represents the deformation of the surface of the vault that resulted from the loads to which it was subject over the course of time. Thus, in the attempt to formulate a hypothetical geometrical model of the surface as it was originally conceived by Mansart, certain criteria were established for interpreting the deformity:

- the curves interpreted as series of concentric arches identify a surface of revolution; geometrical elements characterised by very slight curvatures are interpreted as straight lines (curved as a result of deformation);
- curves classified as conics represent sections of quadric surfaces.

As a result of the analysis based on these criteria, the curves that were determinant for the creation of the model in question (that is, the directrixes and the generatrices) were identified (figs. 21, 22). These curves were derived when horizontal and vertical sections of the point cloud were studied individually from the point of view of geometry.

The intrados as a whole is comprised of three types of surfaces:

- surfaces of revolution (end portions of the small vault);
- quadric surfaces: extruded solids, where the generatrices are circular segments and the directrix is a horizontal line (central portions of the small and large vaults, central groin between the two vaults, lunettes over the entral portals); ellipsoids (end portions of the large vault);

- 2-rail sweep surfaces with one directrix (double corner lunettes, central lunettes, triangular connecting surface in the small vault).

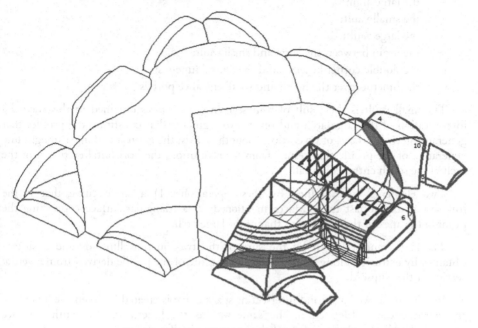

Fig. 21. Drawing of the theorical geometry model

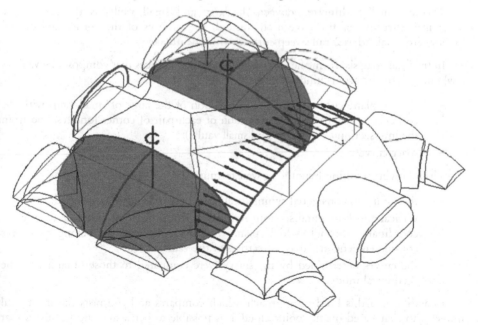

Fig. 22. Drawing of the theorical geometry model

The definitive geometric model is composed of various parts:

- the large arch;
- the small vault;
- the large vault;
- the groin between the large and small vault;
- the double corner lunettes and the central lunettes;
- the lunettes over the north and south entrance portals.

The small vault is the result of four segments of surfaces obtained in this way: the identification of the large arch and on it, two segments that constitute the profiles that generate the two surfaces of revolution about the $z$-axis; the extrusion along a straight line generatirix of the profile that results from the revolution; the horizontal extrusion of the arch of the south entrance portal.

Two elements are obtained by these operations: 1) a space curve that is the intersection of the last two surfaces mentioned; 2) a triangular surface connecting the point of the apex of that curve and that of the large arch.

The large vault is given by the sum of the two quarter-ellipsoids and a surface obtained by extrusion along a horizontal straight line of the profile derived from a vertial section of the ellipsoid.

In this case as well a second intersecting space curve is created between the lunette of the north entrance (derived in the same way as the lunette of the south entrance mentioned earlier) and the surface of the large vault itself.

The intermediate lunette between the large and small vaults is the result of a horizontal extrusion of the large arch. Instead, the surfaces of the corner and central lunettes are obtained as 2-rail sweep surfaces with one directrix.

In the final analysis, from the intersection of all the surfaces that compose the vault as a whole, we obtain various kinds of edges:

- vertical plane curves (from the intersection of the large or small vault with the corner lunettes, from the intersection of each pair of corner lunettes, and from the intersection of the large and small vaults);
- space curves.

The following working hypotheses were formuated:

- the vault was considered symmetrical, thus only one part was analysed;
- vertical and horizontal sections were taken from the point cloud, at exactly the coordinates through which planes that were believed to belong to curves considered generatrices or directrixes of the vault passed;
- the curves determined by the survey were compared to those obtained by the theoretical model.

In addition to this kind of approach, which compares and contrasts the theoretical model with that based on the point cloud, it is possible to perform critical analyses that originate in the geometrical study made of each curve. To this end, the curves of the vault can be grouped into the following categories: the diagonal curves that separate the large and small vault from the corner and central lunettes; the curves of intersection between each pair of corner lunettes; the curves on the east and west walls; the curves on the north and south entrance portals; the curves adjacent to the north and south entrance portals. All of these curves are polycentric, constituted of from two to five circular segments. The geometric methods we used to draw the curves were:

- the method of Abraham Bosse, which is equivalent to applying Archimedes' lemma 3;
- the method of Christiaan Huygens.

Bosse's method is valid for the construction of arches with two or three centres that are symmetrical with respect to the vertical axis. Huygens's method is valid for any number of centres and can be used for both symmetrical and asymmetrical constructions. It was used for the construction of curves with four or five centres. In light of this it can be stated that the diagonal curves belonging to the first category have five centres, symmetrical with respect to the vertical axis, drawn using the method of Huygens's, beginning with a semi-decagon.

Two of the three curves on the east and west walls have three centres, symmetrical with respect to the vertical axis. These were drawn using Bosse's method (that is, by applying Archimedes' third lemma). The last curve, the central one, showed different characteristics: it has four centres, is asymmetrical, and was constructed using Huygen's method.

The curves of intersection between each pair of corner lunettes have two centres, and were drawn using Bosse's method.

The arches of the entrance portals consist in a polycentric curve and two line segments. In particular, the curve of the north entrance has three centres, is symmetrical with respect to the vertical axis, and was drawn using Huygens's method; however, it was not based on a regular polygon, but rather on one of angles 65°, 25°, 25°, 65°.

In contrast, the curve of the south entrance has five centres, is symmetrical with respect to the vertical axis, and was also constructed by Huygens's method, but beginning from the half-decagon.

The curve adjacent to the north entrance has five centres, is symmetrical, and was constructed using Huygen's method, based on the semi-decagon, whereas the curve adjacent to the south entrance has four centres and is not symmetrical, but rather corresponds to angles 45°, 45°, 32°, 58°.

The large arch is a case all of its own. It is the principal element of the vault and can be looked at from three points of view: compositional, constructive and geometric. In terms of composition it defines the space covered by the two vaults. In terms of construction it supports the loads that derive partly from the large vault, and partly from the small. Geometrically it is a three-centred curve.

With regard to the large vault, the two vertical generatrix sections are segments of a single oval. Instead, with regard to the small vault, a series of horizontal and vertical sections taken from the point cloud made it possible to obtain useful information:

- the oblique side of the connecting triangular surface turns out to be parallel to the straight line that is the projection in plan of the diagonal arch of the southeast lunette;
- the small vault is comprised of three parts, of which two are surfaces of revolution and one of extrusion.

Having modelled the surface of the intradox of the vault, the next step was to draw the lines that make up the stereotomic pattern. Particular attention was given to the longitudinal lines that provide the design of the intrados with its continuity, in accordance with Mansart's aesthetic objectives.

The reconstruction of the pattern began with the identification of the blocks of the large arch, using the studies made available to us by Luc Tamboréro, carried out in 1996 on the small vault. Horizontal planes were made to pass through the points identified by the joints of the blocks. The sections cut through the small and large vaults by these planes generated the pattern of the lines on which the two vaults were based.

A different method was used to discover the device used for the double corner lunettes and the central lunettes. In this case the process used was that of the rectification of photographic images of each arch on the wall, from which the pattern of the joints was obtained. Connecting the points of the joints thus obtained with those deriving from the planes of the horizontal sections made it possible to identify the stereotomic pattern.

As we said earlier, the vault was not conceived as an assembly of several surfaces, but rather as a single, complex structure. The unitary nature of the vault is due to the presence of the lines that mask the actual surfaces of the intrados, making those parts simpler and thus capable of being related to cylinders or spheres.

### A different analysis by Luc Tamboréro (Compagnon du Devoir)

A separate analysis of the vault was made by Luc Tamboréro, nicknamed *Persévérance* of Arles, who, in 2001, built a stone model of the small vault at a scale of 1:5, in accordance with the principles of seventeenth-century stereotomy; it was the final work of his studies with the *Compagnons du Devoir*.

The working method adopted was principally based on historic study of the archives, a precise survey and the creation of a model on a scale of 1:5, in Fontvieille stone.

The construction of the vault was based on three points:

- the governing lay out;
- the dual role of the plan view;
- stereotomy and its methods.

### The governing layout

Two interpretative frameworks are superimposed on the setting up of the vault. The first one is linked to the module of the *canne*. For instance (fig. 23), the base square is 8 *cannes* (16.37 m) long on the side, and the columns for the north and south walls are regularly spaced with regard to this unit of measurement (fig. 24).

The second framework concerns the construction of the regular pentagon using straightedge and compass whose opening is set only once and remains fixed in the course of the construction, given by Dürer (see figs. 23 to 25).

This makes it possible to determine the positions of the east and west columns, of the key stones and of the centering point, as well as the position of the penetrations and even the rises of the arches, as we will demonstrate below. The second implementations are, of course, incommensurate with the *canne*.

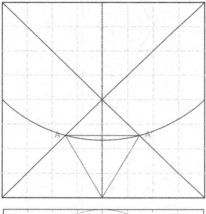

Fig. 23. Drawing of a circle and semi-circle, intersecting according to a bowstring equal to their common radius. The point A and A′ will be the horizontal projections of he keys of the large vault semidomes

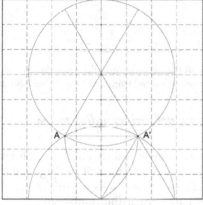

Fig. 24. Drawing of a circle and semi-circle, intersecting according to a bowstring equal to their common radius. The point A and A′ will be the horizontal projections of the keys of the large vault semidomes

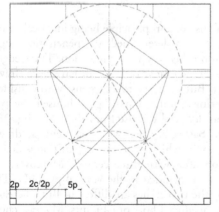

Fig. 25. Construction of the regular pentagon using straightedge and compass whose opening is set only once and remains fixed in the course of the construction, given by Dürer, (the side of the pentagon is the radius of the circle of fig. 23)

The drawing of the North-East and North-West double lunettes clearly confirms the existence of these two frameworks. Indeed, the axial edge, for example, of the North-East lunette joins midpoint 2 of the penetration edge with point 1 (see fig. 28). The drawing is based partly on the modular drawing and partly on the pentagonal drawing, is therefore not orthogonal to the penetration edge. This remark proves that the governing drawing has been scrupulously followed.

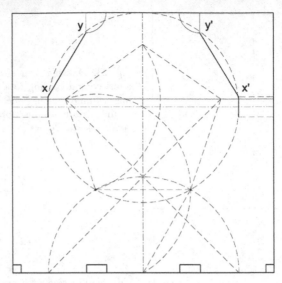

Fig. 26. The two segments x,y and x'y' are not layout of the penetrations of the two double lunettes in the semidomes of the small vault. This drawing was obtained from a modular North-South layout and from a pentagonal East-West layout

## The dual role of the plan view and the volume defined according to the penetrations

One of the principles of constructing the drawing apparently lies in the establishment of the horizontal projections of the penetrations. The penetrations of the large and small vault with the double lunettes are flat; that of the two main vaults is partially flat; those of the entrance lunettes with the principal vaults are circular (horizontal projection) (see fig. 29).

We chose to work on the small vault, its construction principle being identical to the large one. Once the plan implementation had been determined, the penetration edges between the vault and the lunettes were drawn on this very same projection. These edges, which are basket-handles curves with five centers whose span is given, have been drawn following the method of Christiaan Huygens. But contrary to Huygens, who considers the rise as given and chooses the first centering point, Mansart gave himself the two centering points (one of which is point a) and deduced the rise from those (see fig. 28). All plan, penetrations or outline arches are basket-handles with five centers, drawn according to the same method. Only the large arch which support the partition wall and on which the two vaults rest, is a connection arch with three centers. Moreover, the global drawing is based on a template shape arch included in the vertical plan bH. The distance bH being unequal to the distance bK, the two arches are different. But whereas a regular distortion from bK to bH would have produced horizontal joints, Mansart chose two different connecting curves which generate the bending of the intrados.

The order of the drawing and of the construction is therefore as follows:

- the periphery arches and the double lunettes;
- the large arch;
- the small vault with the large entrance lunette;
- the large vault.

The archives also confirm this order.

The rest of the drawing (see figs. 27 and 28) will be obtained by lines between the horizontal projection and the points, on this view of the different arches.

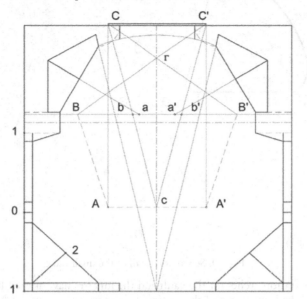

Fig. 27. The segment Br and B'r of the pentagon are extended until they intersect (C and C') with straight line of the spacing AA'. The segment CC' is located at 10 cm above the base square according to the regulator drawing, 3cm according to the tracing; taking into account the encounter angle and the number of previous operations, the mistake in minor. From c, in the middle of AA', we draw cC and cC' with cut BB' respectively at h and h''. The point a and b will be the centering points of future drawing. The point 0, which positions a pair of columns, is the middle of segment 1,1' and does not belong to the straight line AA'.

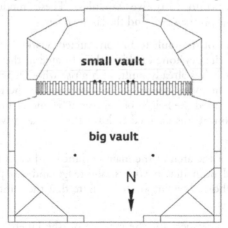

Fig. 28. Schematic of the vault of the vestibule with the subdivision of small and big vaults by the great arch

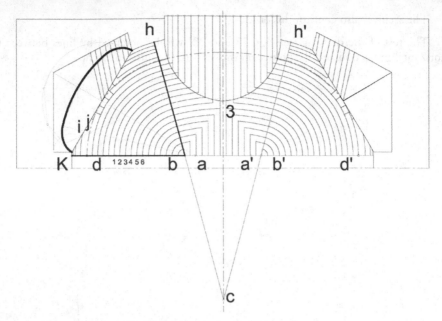

Fig. 29. Schematic detail of the small vault

As the penetration arches for the vault and the lunettes have already been determined, the drawing of the large arch and the drawing of the template shape arch bH are deduced from them (and not the contrary). The five centering points a, a', b, b', c, enable one first of all to draw the curve (d-d'), tangent to the lunettes. Center a is used for transferring the shape of arch I to the cross wall; center b is used for transferring the shape of the arch I to the template arch bH. It is the distortion of the crosswall springing curves which confirms that the order which is followed is indeed the one which has been described. This arch is composed of two distortion curves and a connecting arch which is the only one divided regularly into forty-five modules. These modules will generate the disposition of the joints on the small and the large arches.

As the first element of the vault to be constructed, work on the partition arch began on 24 October 1674; the keystone of this arch is located at the same level as the top of the large vault. The arch was already built, with a rise whose length is one *canne* and one and a half *pan* (243 cm) when the model was received at the beginning of November from Paris. Mansart wanted the height of the rise to be one and half *pan* less (204.6 cm). After a quick discussion, it was decided to leave the arch as it was and to carry on with the work.

The construction of the arch of the main entrance and of its penetration in the small oval vault, which had been drawn with straightedge and compass, followed the same procedure described above. It is important to note that the joint lines do not turn back on the penetration edge.

This fact shows that respect for the module on the lunette, as well as the aesthetic concern about the penetration curve, outweighed the definition of the volume, which his was therefore not defined. The concurrence of this drawing with the survey also provides proof of this hypothesis.

## The stereotomy and its methods

A number of different cutting techniques were used for the construction of this vault.

The first two rows, which are part of the outside wall (and which therefore would have been constructed first) are corbelled and therefore have been square cut. The two semidomes of the large vault, which follow surfaces of revolution, were probably cut by panel according to the method of truncated cones, which was quite traditional at the time. However, on the small vault, the semidomes are not regular, as mentioned earlier.

It is likely that the stones of the six rows of evolution were square cut.

It is also probable that, in view of the good implementation of the joint curves, the vault was constructed on a *veau*, the curve of an arch with a plastered surface where the joint curves were traced.

However, as far as the stones of the lunettes are concerned, a square cut seems impossible for the following reasons:

- the archives mention the fact that Mansart "*baillera le trait à celui qui le conduira*" (will teach those who carry out the work what to do) and Peytret spent approximately one month receiving "*les instructions modèles et panneaux pour lesdits bâtiment et voûtes*" (the instructions for the model and panels for the building and vaults). To be sure, Peytret did not need a month of apprenticeship with the master in order to use the square cut technique;

- A model was made out of wood for the large lunette, which also suggests the use of a more complex method than the square cut;

- We know that the square cut was not a method advocated by the Académie and Mansart himself would have been the last to have campaigned for its use;

- Last but not least, the square cut uses a huge quantity of stone at the level of the haunch of the vault, especially with the curved penetrations and for a vault of at least 50 cm thick.

The square cut is therefore excluded for the lunettes. But a traditional cutting by panel does not seem appropriate either. Indeed, all the intrados are warped, which eliminates the (numerous) drawings out of a flat intrados panel.

Moreover, the panel method requires different panels for each stone, given the irregularity of the stones on a same row and in between rows. This time consuming drawing does not seem to be compatible with either the rapidity with which the work progressed of the work or with the archives, which mention a limited number of patterns for the panels.

Given the complexity of this kind of vault, the traditional stereotomic methods are inadequate. We therefore maintain that the stonecutting method used was that called *à la sauterelle* (the name *sauterelle* comes from the name given by the workers to the 'false square'). Used in carpentry, this was a quick method for cutting arch stones, using simple angle transfers and without the need for drawing all the panels. The principle is as follows: to determine a polyhedron with six sides, it is necessary, for each corner, to know the angles of the edges which converge to it. Here one of the sides is warped but the knowledge of those angles is sufficient for the cutting. Starting from one of the joint plans, two segments of straight lines of the intrados – non coplanar segments – will make it possible to adjust the intrados during the cutting process. Figs. 30, 31 and 32 show respectively the drawing of the angles needed for the construction.

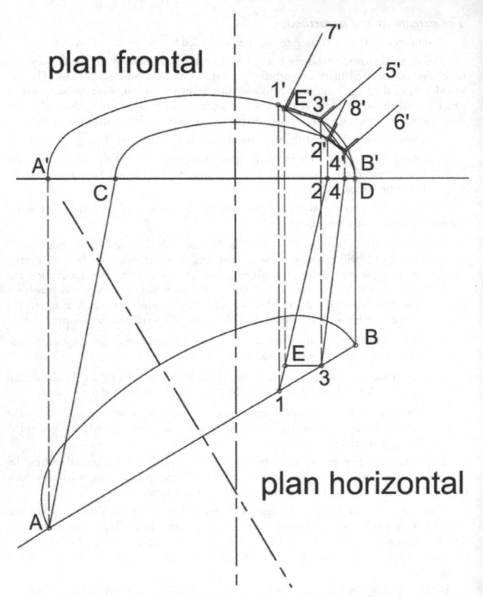

Fig. 30. Head angles. The lunette is defined by the arches AB and CD, whose elevations are projected down to the plan. The arch AB projects itself into frontal A'B'. Any given soffit is defined by its plan 1-2-3-4. In the terminology used in French carpentry, the frontal plan is "*la herse*" and the horizontal plan is "*le plan*". Segment 3E is a "*chevron d'emprunt*", that is, a construction line which is not the outline of the piece but a marker for its execution, and is parallel to the frontal plan. The straight line 6'-4' goes through one of the centers of arch CD. It makes it possible to define the joints of the voussoir: it is sufficient to consider a parallel line 3'-5' through the frontal plan 3E. In the same way, we can define the other side of the joint 8'-2'-1'-7'. The angles needed for cutting the stones appear on the frontal projection and will be measured using a *sauterelle*, an two-side adjustable template

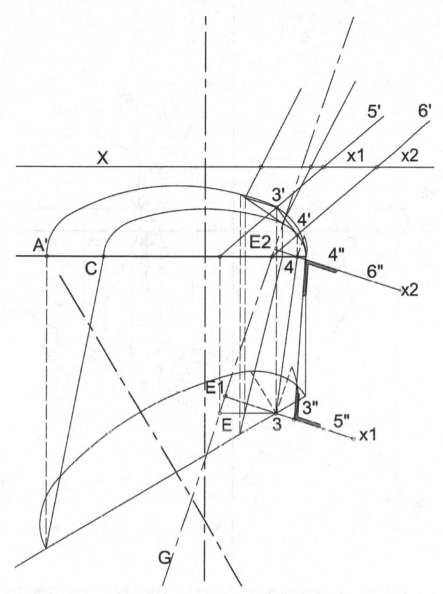

Fig. 31. Joint angles. The plan of the joints (6'-4'-3'-5') is projected down to the horizontal plan, and the angles of the bed is deduced from that. Also indicated in the figure is the horizontal plane X. Taking the distances 3'-x1 and 4'-x2 allows you to locate X on the stone, and to square cut the penetration

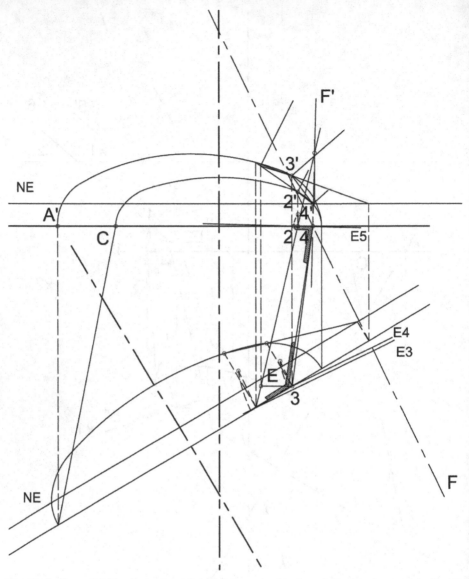

Fig. 32. Soffit angles. To obtain the real size of the angle, the plane defined by two straight lines 3-4 and 3-E is rotated down around a horizontal hinge (for instance, the one which goes through point 3). In the same way, we can determine the angle for the two straight lines 3-4 and 2-4 thanks to a hinge going through 4. A length transferred onto segments 3-E and 2-4 determine the second edge of the soffit

The first French publication that dealt with *sauterelle* dates to the eighteenth century, but the method itself dates to before that, because it is described in the 1619 manuscript by Diego Lopez de Arenas. This Spanish manuscript provides us with information regarding the origin of the *sauterelle*, a method that appears to have been developed to deal with the complexity of geometric ornamentation inspired by Islamic art and architecture.

Another hypothesis regarding knowledge of this method derives from the fact that Girard Desargues used a method of double projection in order to take from the drawing the true dimension of the angles that the planes make with one another. Even though the arrangement and angles used by Desargues in his demonstrations are not anything like those of the *sauterelle*, the tools of projection are the same.

It is possible that Mansart came to learn about this method through studies of the works of Desargues during the three years he spent in the seminaries in Versailles, until 1673, before his journey to the south.

For the penetration arch stones, the X plan is used (see fig. 31) on which the panel of the horizontal projection is applied. The penetration can be square cut since the radial joints of the small vault are vertical.

## Mansart's spatial vision

Thoroughly knowledgeable in geometry and optics, Mansart knew that the human eye was not capable of perceiving whether the geometry of a surface was exact or not. Because he would not have been able to control the intersections of the portions of the vault, and thus the joints of the courses in space, he devised a solution: tracing the lines of the courses in plan and projecting them on the surfaces of the space would have fooled the human eye. For example, the arches of intersection between the small vault and the side lunettes, and between the large vaults and the side lunettes are exactly plane arches (represented in plan by line segments). The intersections between the cylindrical surface of the large arch and the cylindrical surface that connects the two semidomes of the large vault are also represented in plan as line segments, even though the two cylinders do not have the same radius of curvature. Thus in all probability the drawing of the courses were laid out in plan, and successively the blocks themselves were studied in order to confer on the structure the vision of the whole that the architect wanted to achieve. Obviously, there were some points in which it was not possible to create this kind of block, and so it was decided to draw the lines of the courses.

## Conclusions

The daring construction of the vaulted space of the *Hôtel de ville* had no successive echoes. The spatial studies that linked progress in the discipline of stereotomy to the quality of architectural space had by that time come to an end.

Jules Hardouin-Mansart possessed the necessary knowledge of geometrical technique and structural mechanics to be able to create an aesthetically-pleasing architecture. His work represents the culmination of stereotomic construction. The studies presented here were aimed at filling in the gaps in our knowledge of the vault by, for example, carrying out an accurate survey; but they are also useful for advancing our knowledge of the discipline of stereotomy.

# References

ANDROUET DU CERCEAU, Jacques. 1988. *Le plus excellents bastiments de France* (1576-1579), Paris: Editions Sand.

BARRIELLE, Jean-Francois. 1982. *Le style Louis XIV*. Paris: Editions Flammarion.

BEAUPRE, Charles Mauricheau. 1948. *Versailles, Les Documents d'Art*. Munich: Editions Imprimerie Nationale.

BENOIT, Francois. 1935. Histoire municipale d'Arles. In *Encyclopédie départementale*, Bouche du Rhône.

BOIDI, Giuseppe. 1914. *I cinque ordini del Vignola, da Manuale di disegno architettonico*. Turin: Editions Momo.

BOSSE, Abraham. 1643. *La pratique du trait a preuve de Mr. Desargues Lyonnois, pour la Coupe des Pierres en l'Architecture*. Paris: Editions Imprimerie de P. Des-Hayes.

BOUGARD, Jean-Francois. 2007. *Les Maîtres Bâtisseurs ou la Science des Nombres*. Paris: Editions Mosaique.

BOURGET, Paul. - CATTANI, G. 1956. *Jules Hardouin Mansart*. Paris: Editions Vincent, Freal & C.

BOYER, Jean. 1969. *Jules Hardouin Mansart et l'hotel de ville d'Arles*. Arles: Ville d'Arles.

CARDONE, Vito. 1996. *Gaspard Monge, Scienziato della rivoluzione*. Naples: Editions Cuen.

CHARVET, Leon. 1898. *L'hotel de ville d'Arles, Réunion des Sociétés des Beaux-arts*. Paris.

CHARVET, Leon. 1870. *Les Royers de La Valfenière*. Lyon.

CHASTEL, Andrè. 2006. *L'Art Français ancient régime 1620-1775*. Paris: Editions Flammarion.

CHITHAM, Robert. 1987., *Gli ordini classici in architettura*. Milan: Editions Hoepli.

DE LA HIRE, Philippe. 1687-1690. *Traité de la coupe des pierres*. Paris: Bibliotheque de l'Institut de France, Ms. 1596.

DE LA RUE, Jean Baptiste. 1728. *Traité de la coupe des pierres*. Paris: Imprimerie Royale.

DELORME, Philibert. 1567. *Le premier tome de l'architecture*. Paris: Editions Morel.

FALLACARA, Giuseppe. 2007. *Verso una progettazione stereotomia*. Bari: Editions Aracne.

FREZIER, Amédée-François. 1737-1739. *La théorie et la pratique de la coupe des pierres et des bois ... ou traité de stéréotomie a l'usage de l'architecture*. Strasbourg-Paris: Jean Daniel Doulsseker-L. H. Guerin.

GIORDANO, Andrea. 1999. *Cupole, volte e altre superfici. La genesi e la forma*. Turin: Utet.

GUARINI, Guarino. 1968. *Architettura civile* (1737). Milan: Il Polifilo.

HEIJMANS, Marc, J. M. ROUQUETTE and C. SINTES. 2006. *Arles antique, guide archéologiques de la France*. Paris: Editions du patrimoine.

KOENIG, Giovanni Klaus, et al. 1993. *Tecnologia delle costruzioni*. Florence: Editions Le Monnier.

MASSOT, Jean Luc. 1992. *Architecture e Décoration du XVI au XIX siècle*. Vol. 2 of *L'art de Restaurer en provence*. Salerno: Editions Edisud.

MICHEL, Alain. 1980. *Au Pays d'Arles. Planches hors texte*. Arles: Editions Arthaud.

MIGLIARI, Riccardo. 2000. *Il disegno e la pietra. Rilievo e Stereotipia*. Rome: Editions Cangemi.

MONGE, Gaspard. 1900. *Géométrie Descriptive* (1798-1799). Paris: Editions Baudouin Imprimeur du Corps Legislatifs et de l'Institut National.

MOTTANA, Annibale, Rodolfo CRESPI and Giuseppe LIBORIO. 1985. *Minerali e rocce*. Milan: Editions Arnoldo Mondadori.

PEROUSE DE MONTCLOS, Jean Marie. 1982. *L'architecture à la française*. Paris: Editions Picard.

———. 1983. *La voûte de l'Hôtel de ville d'Arles est-elle le produit de la tradition locale ou une importation parisienne? Travaux et colloques de l'Institut d'art*. Publications de l'Université de Provence.

PEZET, Maurice. 1982. *Les belles heures du pays d'Arles*. Marseille: Editions Jeanne Laffitte.

RONDELET, Jean. 2005. *Trattato teorico e pratico dell'arte di edificare* (1802-1817). Rome: Editions Librerie Dedalo.

SAULE, Beatrix. - MEYER, D. 2007. *Versailles. Guida per la visita*. Versailles: Editions Art Lys.

TAMBORÉRO, Luc and Joël SAKAROVITCH. 2003. *The vault of Arles City Hall: a carpentry outline for a stone vault?* Pp. 1899-1907 in vol. III of *Proceedings of the First International Congress on Construction History*, Madrid, 20-24 January 2003, Santiago Huerta, ed. Madrid: Instituto Juan de Herrera, Escuela Técnica Superior de Arquitectura.

TOMAN, Rolf. 1998. *L'art du Baroque Architecture. Sculpture. Peinture.* Paris: Editions Konemann.

## About the authors

Giuseppe Fallacara is an architect and researcher at the Faculty of Architecture at the Polytechnic of Bari. Since 2005 he has conducted experiments in stereotomy with the creation of construction elements in stone. Examples are: Escalier Ridolfi, an entry portal for the Venice Biennale (a variation of the Abeille vault), Alexandros obelisk, pre-stressed stone arch built in Brignoles, Toulon (France), arch leaf in Parabita, Lecce (Italy), free-standing stereotomic wall hangings, etc. His publications include: *Verso una progettazione stereotomica* (Rome: Aracne, 2007), *Plaited Stereotomy. Stone Vaults for the Modern World* (with R. Etlin and L. Tamboréro; Rome: Aracne, 2008) *and Domus Benedictae: villa unifamiliare a Corato* (Collana Archinauti, 27; Bari: PoliBa Press, 2010).

Fiore Resta is an architect. He holds a Ph.D. from the Faculty of Architecture of the Politecnico di Bari, in Architectural Design for the countries of the Mediterranean. His thesis is entitled "Spaces of vaults and domes in the architecture of Jules Hardouin-Mansart."

Nicoletta Spallucci is an architect. She holds a Ph.D. from the Faculty of Architecture of the Politecnico di Bari, in Architectural Design for the countries of the Mediterranean. Her thesis is entitled "Vaulted roofs and domes of the Cistercian abbeys restored by Viollet-Le-Duc."

Luc Tamboréro has been a master stonemason since 1995. He is currently a Ph.D. candidate in Architecture at the laboratory Geometry, Structure, Architecture of the *Ecole d'Architecture Paris-Malaquais* under the advisement of Joël Sakarovitch. He studied materials and techniques in France under the terms of *Compagnon du Devoir.* For several years he and his company, Mecastone, have worked on construction sites of both new buildings and restoration projects in different countries. He has produced prototypes for a new building in solid stone.

Rocío Carvajal
Alcaide

C/Costa Brava n°20
esc 3 piso 7°I
28034 Madrid SPAIN
rcarvajal.eps@ceu.es

Research

# Stairs in the Architecture Notebook of Juan de Portor y Castro: An Insight into Ruled Surfaces

Keywords: Historical treatise,
Juan de Portor y Castro, stair
design, paraboloid, descriptive
geometry, stereotomy

**Abstract.** Historic treatises on stonecutting of the sixteenth and seventeenth centuries contributed to progress in the study and understanding of properties of the different surfaces and the intersections between them. One of the topics that contributed most towards the development of knowledge of the geometry of surfaces was the practical use of ruled surfaces. Defining the geometry of a piece of stone by surfaces that can be cut using a ruler as a guide and a check has always been considered very interesting in the field of stonecutting. This present study focuses on stairs in two Spanish treatises to illustrate how warped surfaces were treated.

## Introduction

The notebook about architecture by Juan de Portor y Castro (BNE Ms. 9114) is one of the most interesting texts in terms of Spanish stonework in the eighteenth century. It compiles numerous and varied stonecutting models, some of which are original and some which were copied from other printed treatises available in Spain at the time, such as *Arte y uso de Arquitectura* by the Augustinian monk Fray Lorenzo de San Nicolás [1639-1664], and the *Compendio Matematico* by the priest Thomás Vicente Tosca [1707-1715]. Portor's manuscript, written at a relatively late date (1708), shows how the knowledge on stonecutting was transmitted in Spain. Similar documents on stonework were printed in France as early as the sixteenth century [de L'Orme 1567]. Again in France, plenty of treatises on stonecutting appeared throughout the seventeenth century, culminating with the work by Amédée-François Frèzier, *Traité de stéréotomie* [1737-1739]. In the meantime, knowledge obtained empirically was transmitted, even between different geographic areas, through the aforementioned manuscripts, which were copied and passed on between stonemasons. By reading of Portor's notebook, we can identify different dates that appear on it, showing the notebook was written over an interval of time between 1708 and 1719 when the author moved between Granada and Galicia [Taín 1998: 67, 269]. The degree of interest raised by these manuscripts' contents, and the quality of the graphics they offer are not at all inferior to those of printed works, as proved by the examples of Vandelvira (1575-1580) and Ginés Martínez de Aranda (ca.1600).

These notebooks and treatises comprise a large and varied number of examples of stone bonds. They show a broad range of solutions to the various problems that could potentially arise in the practice of stone cutting. This contributed to progress in the study and understanding of the properties of the different surfaces and the intersections between them. One of the topics that contributed most towards the development of knowledge of the geometry of surfaces is the practical use of ruled surfaces. Defining the geometry of a piece of stone by surfaces that can be cut using a ruler as a guide and a check has always been considered very interesting in the field of stonecutting. The empirical knowledge and practical application of the properties of these surfaces is crucial when it comes to defining how the building elements are broken down into individual stones.

In the case of warped surfaces, besides splayed elements – which are, due to their complexity, the most representative elements that use this kind of surfaces–, it is worth focusing on the study of stairs.

The many examples of stairs analysed in Portor's notebook and the detail of the descriptions are not rivalled by those in Vandelvira's manuscript, in spite of the importance of the variety of this manuscript's content.

The definition of stairs does not refer in a general way to all possible types of stairs, but to those comprised of straight flights called cloister stairs. This kind of stairs are usually placed in one of the cloister's walls. They have a square or rectangular layout and its flights run along three of the four planes that define the rectangular box where the stair is located. Portor, like Vandelvira and Aranda, distinguishes between this kind of stairs and spiral stairs.

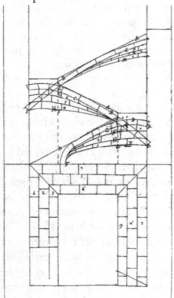 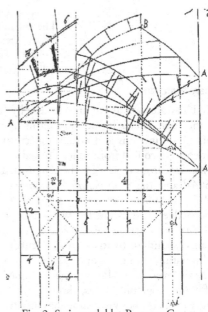

Fig. 1. Stair model by Vandelvira
[ETSAM Ms. R31: fol. 58r]

Fig. 2. Stair model by Portor y Castro
[BNE Ms. 9114: fol. 26r]

The graphic representations in the manuscripts about stonework of these stairs with square or rectangular layout show a number of particular features that require an explanation in order to better understand the didactic intention of the models given for the stairs. This is different from the concept of coordinated projections found in the double orthogonal projection [Rabasa 2000: 337]. The elevation found accompanying the plan view in Portor's manuscript could be interpreted as a dog-leg stair, when in fact, the plan view shows that it is comprised of three flights that run along the planes that enclose the staircase. Vandelvira presents an elevation of each of the flights placed next to each other, showing that the plan and elevation views are not yet two coordinated views (fig. 1). Portor goes a step further in the graphic representation of these stairs and alongside these consecutive elevations of the flights he draws auxiliary views of the joints perpendicular to the strings, offering front views of the planes containing these joints (fig. 2). Thus the same drawing shows all the joints that define each of the voussoirs in true shape, thanks to representing the front view of all the planes that contain the joints [Palacios 1990: 129].

Strings in these stairs form smooth, harmonious intersections. They are also a key example of the skill demonstrated by stonemasons when constructing ruled surfaces. The design of these stairs could be accomplished by either using curved or straight strings. Once the outline of each flight of the stair was defined, the intrados surface of the string could be a vaulted or a plane surface. Thus, Portor's models are divided into two large groups. This article will focus on the analysis of stairs with straight flights, therefore defining a ruled surface.

This type of cloister stairs was broadly used throughout the Spanish Renaissance and many examples are preserved. Some of these examples were cited by Portor to illustrate the models described in his notebook and discussed below.

Portor's notebook compiles models for seven types of stairs: square and rectangular ruled stairs on flying buttresses and pendentives; square splayed warped stairs and square warped splayed stairs with truss. These models are built with transverse, longitudinal or concentric courses.

Pendentives and spiral stairs are shown on the back of the pages featuring the stairs. They appear in the notebook in the same order as in the table on page 63, which lists the models given in pages 1 to 24. On the back of the last stair model, the skew square stair with levelled landings built with longitudinal courses, the corner pendentive in round tower at floor level is shown. It seems likely that all the models Portor intended to include in the notebook he did in fact produce and have been preserved. This is not the case for the manuscript by Aranda, although it seems clear that he had planned to write a section. He wrote:

> For greater clarity, I have divided them [my writing] into five parts. In the first part I will deal with difficult arches; in the second, with splayed arches and doorways; in the third, with stairs, including spiral stairs; in the forth, with pendentives and vaults, and in the fifth, with chapels and chamfered corners [Martínez de Aranda, ca. 1600: proem (my translation)].

However, no stair designs are found in the manuscript. Only two designs appear in Vandelvira's manuscript: "truss stair with smooth transitions" and "ruled stair with abrupt transitions".

In the design of a cloister stair, the flights span between inclined planes and are articulated by two landings, and the definition of the surface described by the strings between its two edges is crucial. The solution to the problem of the intersection of flights at the landings depends on the nature of these intrados surfaces. Portor describes various possibilities for obtaining a successful transition between flights. These possibilities can be divided into two groups: those that generate surfaces limited by curved lines, and those that have strings defined by straight lines. The latter generates warped ruled surfaces between the straight lines that delimit the different flights.

Such warped ruled surfaces will be hyperbolic paraboloids and their most significant properties will be described in these models. The joints between voussoirs and the intersections between flights are obtained when these surfaces are cut by planes. This reflects the master builders' deep knowledge of the geometry of the different surfaces and, in this case, of *engauchidas* (warped) surfaces, as Portor calls them. Master builders gained this knowledge in an empirical way and it was later synthesised in stereotomy treatises.

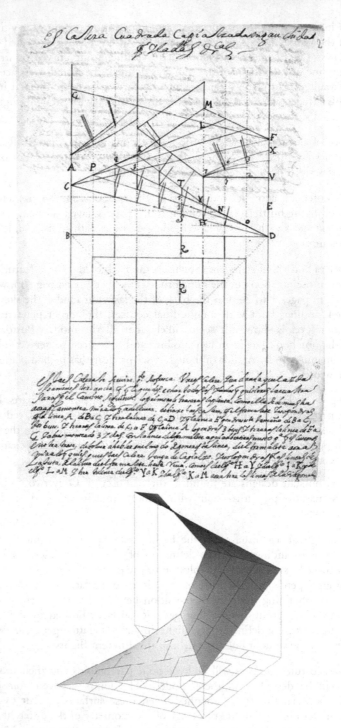

Fig. 3. Square splayed warped stair built with longitudinal courses
[Portor, BNE Ms. 9114: fol. 23r]

## Square splayed warped stair built with longitudinal courses

This stair (fig. 3) is not featured in Vandelvira's manuscript, which only provides the opposite case built with transverse courses, that is, courses that are perpendicular to the head line of the strings instead of being parallel. Portor recommends using this stair when the flights must have a substantial width, because it is possible to add as many longitudinal courses as needed.

The support plane for each flight is defined by four lines: the line of intersection between this support plane and the head of the string delimiting the stairwell; the intersection with the facing on which the stair rests; and the two intersections with the previous and successive flights. These four lines delimit the warped quadrilateral ABCD that defines a hyperbolic paraboloid (fig. 4). The horizontal projections of lines AB and CD are parallel to each other. Therefore, the axis of the paraboloid will be a straight horizontal line for the said projections and the direction of one of the plane directors will be that of the vertical planes that contain the projections. Therefore, the longitudinal joints between courses will be generatrices of the surfaces, since they are the intersection of the paraboloid with planes that are parallel to a plane director. The case is not the same for the transverse joints between voussoirs of the same course. These joint lines are generated by the intersection of vertical planes with the paraboloid. The planes also cut the axis of the surface, and therefore the lines will be hyperbolae (fig. 5).

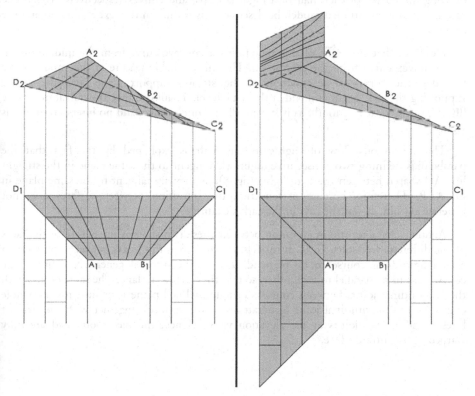

Fig. 4 (left). Paraboloid defined by the warped quadrilateral ABCD
Fig. 5 (right). Longitudinal courses and Transverse joints

In the plan of this stair, Portor draws all these transverse joints as curved lines, although he states that they can be drawn either curved or straight. Furthermore, he differentiates the curvature of each half of a flight because the directrix that goes from the mid-point of the generatrix AB to the mid-point of the generatrix BC defines the change of curvature in the paraboloid. His drawings also show how the curvature of these joints decreases as they come closer to this generatrix. Despite Portor's not knowing the definition of these conics, in the plans for this stair he describes not only the nature of the lines of the joints, differentiating straight lines from curved ones, but also more precise aspects, such as the double curvature of this surface.

### Square skew stair with levelled landings and built with longitudinal courses

Taking the previous model as a reference, in this stair only the height of the points A, D and BC (springing and end of each flight) has been modified. In this case, they are all at the same level so that the lines AD and BC are straight horizontal lines (fig. 6).

A warped quadrilateral is obtained in this case, and this quadrilateral defines each of the strings of this stair. The arrangement of the longitudinal courses in this model is identical to that of the design described above, and therefore these joints will be straight and will define generatrices of the paraboloid (fig. 7). The same happens with the transverse joints between the voussoirs, which are once again curved. Here, instead of drawing the curves of each half of a flight concave and convex respectively, as was the case in the aforementioned model, he describes this feature in the text accompanying the drawing.

He states that the soffit of this stair forms a concave curve from the middle down, and a convex curve from the middle up. The most notable feature of this model is the fact that the curves of intersection of the string's support plane with the facing supporting the stair begin and end at the same level. Thanks to this, the transition of one flight into the next is smooth, as if it were all just one surface, and no intersection line is seen.

The continuous flow of flights in this stair is explained by the fact that the paraboloids defining two consecutive flights are tangent to the same plane in the straight line AD shared between the two flights (fig. 8), as they are tangent to the same plane in D, A, P. Therefore, because of tangency, the transition between surfaces does not generate a groin, as was the case in the earlier example.

Another characteristic in common between these two stairs is the nature of the faces of the different voussoirs. In both models, Portor forces the joints between voussoirs in each longitudinal course to be perpendicular to the internal generatrix defining the course, and to be parallel to each other, so that they define a plane. The same occurs with the longitudinal joints between courses; a single inclined plane is defined for the whole course by the longitudinal joint (generatrices of the surface), together with the straight lines found in the joints at the elevation view. Hence, the pieces obtained are only warped in the intrados face.

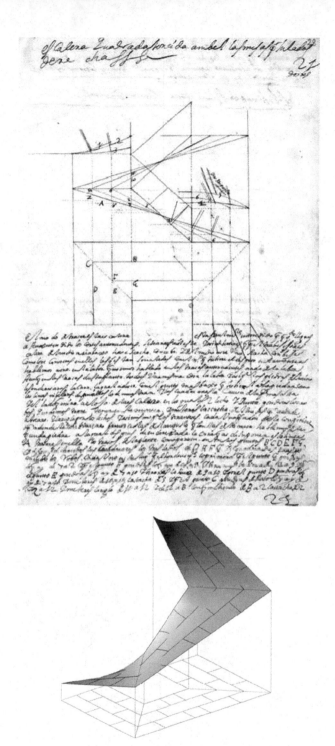

Fig. 6. Skew stair with levelled landings and built with longitudinal courses
[Portor, BNE Ms. 9114: fol. 28r]

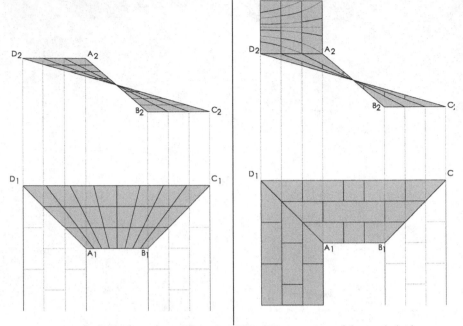

Fig. 7. a, left) Warped quadrilateral ABCD and generatrices of the paraboloid;
b, right) Longitudinal courses and transverse joints

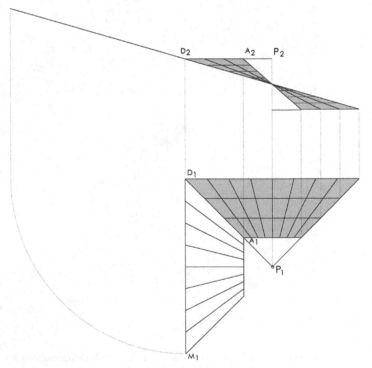

Fig. 8. A, D, P: Points of tangent planes to both paraboloids

Further on, Portor refers again to this model in the *splayed oblique cross vault for a flat stair without treads–ramp* (fig. 9), where he begins the explanation with the following words:

> ...this cut, I believe, is the most intelligent cut that I have given thus far, and in order to understand it I think it is first necessary to try to understand the previous stair with levelled landings, because it is very similar in all aspects. Indeed, the springing lines described by the arches of this vault define a warped plane equal to the plane that supports this stair [Portor 1708: fol. 48r].

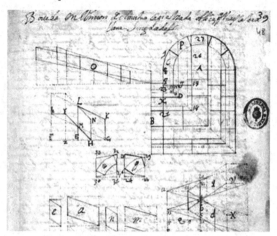

Fig. 9. Splayed oblique cross vault for a flat stair without treads [Portor, BNE Ms. 9114: fol. 48r]

## Warped splayed stair built with transverse courses

The design for the warped splayed stair built with transverse courses (fig. 10) is different from the first we have mentioned in that the arrangement of the courses is different. Here the courses are placed perpendicular to the string. Thus, the collection of ruled stairs becomes more complete with the introduction of a new variable: the direction of the courses.

This stair is also described by Vandelvira and Fray Lorenzo de San Nicolás. Each one of these descriptions begins with praise of its ingenious design. In Vandelvira's manuscript, it is referred to as "ruled stair with smooth transitions" (fig. 11). Vandelvira writes about it: "This stair is the most elegant and artistic I have found because, once they are understood, they are very pleasant, for they defined by ruled surfaces everywhere, as splayed ruled arches" [Vandelvira ca. 1575: fol. 59v]. He illustrates each one of the cases he shows, ruled or curved, with a single example. He develops the option of courses arranged in the transverse direction for the case of the ruled stair with smooth transitions, and the arrangement in "straight" (longitudinal) courses for the case of the curved stair with smooth transition.

Portor, in contrast, analyses all the possible combinations between ruled and curved designs, and the different possible arrangements of the courses. He even shows the case in which the courses are arranged curved between the two lines that delimit the string of the stair, as is the case of the warped square splayed stair with transverse courses with curved courses.

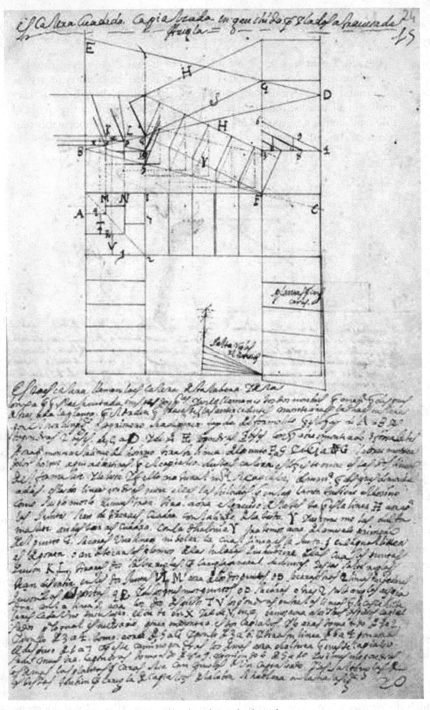

Fig. 10. Square warped splayed stair built with transverse courses
[Portor, BNE Ms. 9114: fol. 24r]

*[handwritten text in old Spanish script, partially legible]*

*y a la linea queba de la. L. e a la L. pues despues de aber acabado. las buestas mas anter*
*en la cultima pues lo queay de la. g. a la. m. y la quarta de la. o. a la. q. a la*
*quarta deliz a la. m.: yesta a la buelta poco... que aun que en la tenta repara con*
*mas de mas buebtas de la menor quite... a esquadra ques en medio de las dar no tal*
*que no... muchas! y la poltera a ble la y. a la m. como de mas buelta la. o.*
*y obre porque a bler y son engas... das e la figura. T. ael rincon abaxo do que*
*estan la piedras... a le contenido... y en pan ella y ence de la menor quita*
*encienda quita anch... y los rincones le tocan en algun estadio tomando guen te en*
*pe punto de cada linea e a la figuedria —*

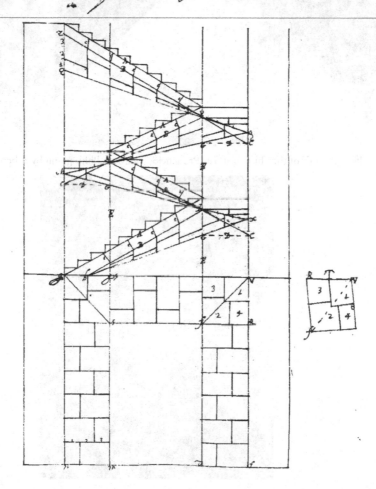

Fig. 11. Ruled stair with smooth transitions
[Vandelvira, ETSAM Ms. R31: fol. 60 r]

The text in Portor's notebook that accompanies the model for this "square warped splayed stair built with transverse courses" starts by citing two significant examples: "The stairs of Talavera and of the Guildhall are said to be this stair because it is constructed in those two parts, and thus it is called these two names" [Portor 1708: fol. 24r] .

Fig. 12. Stair of the church of San Prudencio, Talavera. Photograph by Alberto Sanjurjo

Fig. 13. Stair of the Guildhall, Sevilla. Photograph by Alberto Sanjurjo

Portor is talking about the stair that goes up to the choir of the church of San Prudencio (the old convent of Hieronymites of Santa Catalina) in Talavera de la Reina (fig. 12) and stair of the Guildhall of Sevilla, now the Archivo de Indias (fig. 13). Palacios cites the stair of the Guildhall to illustrate the explanation of Vandelvira's ruled stair with smooth transitions [1990: 182-184].

It can be seen that the stair in the Guildhall faithfully reproduces this stair model. Only insignificant variations are found when compared Portor's drawing. This is not the case of the church of San Prudencio. In this stair, the joints that are normal to the string are not straight, but slightly curved, and the thickness of the head is not constant throughout, but decreases as each flight goes up.

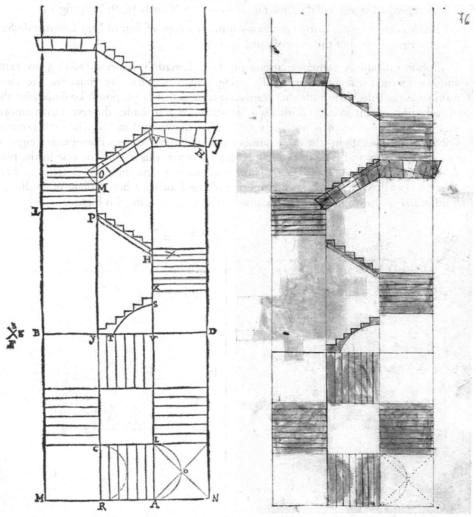

Fig. 14. Fray Lorenzo de San Nicolás [1639: 119]

Fig. 15. [Portor, BNE Ms. 9114: fol. 76]

There is no doubt that both stairs follow the same pattern in their design. However, the models found in the manuscripts and in the stonework treatises are not exact reproductions of real examples, but general schemes that require adaptation to particular conditions of each specific case.

Fray Lorenzo de San Nicolás writes about the same stair in Talavera in his book *Arte y Uso de Arquitectura*:

> After explaining the making of the timber stair, I must deal with the cuts of other stone stairs, making use of the stair found in the convent of Santa Catalina of the Hieronymites in the town of Talavera, later copied in the convent of Uclés of the Military order of Saint James. The stair is so ingenious that I will demonstrate its cuts [San Nicolás 1639: 119] (fig.14).

Besides this design, Portor's notebook features a copy of that of Fray Lorenzo de San Nicolás, near the end of the manuscript (fig. 15).

Portor introduces some variations on Fray Lorenzo's drawing, adding an extra voussoir in the breaking down of the string, which now has five voussoirs, the same number as the ruled stair built with transverse courses that he proposes. To emphasise the introduction of this stone, Portor leaves it unshaded and shades the rest. Furthermore, this distribution of steps is not the same as that drawn by San Nicolás. Fray Lorenzo shows six steps per flight (the same number as Vandelvira), whereas Portor draws eight in two of the flights and nine in those opposite. However, this is only the case in the plan drawings. In the elevation drawing, Portor shows the same number of risers as Fray Lorenzo. Lastly, with respect to the top part of the drawing, where a flight of a different kind of stair is shown, he is not meticulous at all in representing this flight.

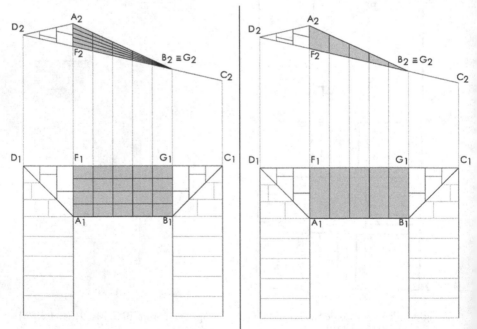

Fig. 16. a, left) Warped quadrilateral ABCD and generatrices of the paraboloid; b, right) Longitudinal courses and transverse joints

Having described the representations of this stair in the others manuscripts, and seen the built examples cited, we can proceed with the detailed study of the model given in the manuscript by Portor. In this notebook, all the different flights of the stair spring from the same level, that is, the intrados edges that define the plane of the string start at the same height but end at different levels, the final point of the head line being always below the final point of the line that runs along the vertical facing delimiting the string. This line extends until it reaches the opposite wall, and at this point it connects with the next flight. This generates in the soffit of a landing an intersection of two triangular planes that are the extension of the two flights that meet at the landing. Therefore, we have an intrados surface defined by a straight section (landing) and a warped section (string) (fig.16). The warped plane defined by each flight is a ruled paraboloid delimited by the warped quadrilateral ABFG, sharing the straight lines AF and BG with the triangular planes that form the landings.

The fact that the horizontal projection of the warped quadrilateral that defines the paraboloid is a rectangle, simplifies the identification of the direction of the surface's plane directors. These planes are vertical and are parallel to the sides of the rectangle.

In the case of the "skew square stair with levelled landings built with longitudinal courses" we saw how the transition between warped surfaces was continuous because the paraboloids that define the strings were tangent along the common generatrix. In the stair we are now dealing with, each flight is comprised of two surfaces, one of which is developable, while the other one is not. Therefore, they cannot be tangent along a generatrix because for the warped surface the tangent planes vary along the said generatrix, whereas on the plane of the landing, the tangent plane is unique. However, in this case, this is not significant. Thanks to the arrangement of the joints in the direction perpendicular to the string, the joints between courses conceal the discontinuity between the two surfaces, which becomes imperceptible.

Thus, each flight simulates a continuous surface, with the only visible groin being the diagonal where the flights meet under the landing.

The breaking down of this stair is obtained using only straight lines. The lines of the joints between courses, as well as of the joints between stones of the same course (in case these were necessary due to the width of the string), are generated by the intersection between the intrados surface and vertical planes parallel to the plane directors of the said surface, and therefore will also be parallel to its axis and will define generatrices.

As in the cases analysed above, Portor once more defines the side faces of the courses using planes, so that the warped surface is only found on the face of the intrados (fig. 17). These planes go through the joints between courses and will intersect the planes of the vertical facing that contains the string forming parallel lines, as explained in the accompanying text:

> Now for the joints, they must be drawn perpendicular to the intrados lines around the stairwell Y, and similarly, the edge where the intrados meets the wall must be perpendicular with the said line Y [Portor, BNE Ms. 9114: fol. 24r].

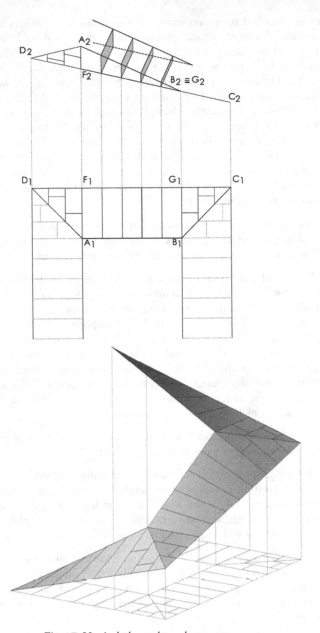

Fig. 17. Vertical planes through courses

The analysis of these models for stairs with straight flights offers an insight into the properties of this type of surfaces, later to be called hyperbolic paraboloids and defined as tri-axial, warped, ruled surfaces with a plane director. The author of the manuscript did not know the name of the surfaces; nor the classifications of ruled surfaces (it was sufficient for him to distinguish between developable and warped); nor terms like "plane director", "asymptotic plane", "axis" or "vertex of the surface", and so forth. However, he did know how to distribute the courses to originate joints that were either straight or curved lines; for double curvatures, he identified where they were concave or convex and could, therefore, correctly identify the different sections that would be generated by a plane on such a warped surface thanks to his empirical knowing. He was capable of proposing different solutions to guarantee a smooth transition between surfaces without knowing the tangential conditions that could be applied. As José Calvo Lopez writes:

> Descriptive Geometry didn't come to life in the silent study of a wise man; it was born surrounded by the dust of masons' guilds and in the heat of artillery battles. Many of the key basic known notions of the discipline, such as the orthogonal projection or how to obtain auxiliary views, were not the result of abstract thinking; on the contrary, they were devised empirically by the builders of the later Gothic and the Renaissance [Calvo 2002: 313].

The non-developable condition was the greatest concern for the authors of these manuscripts; the manuscripts are full of estimates, more or less closely approximated, to try to "extend" a warped surface along a plane. All these thoughts contribute to the idea that descriptive geometry involves the systematization of the teachings that stonemasons passed on and developed in the actual practice of their work on stonecutting.

## References

### Manuscripts

PORTOR Y CASTRO, Juan de (attributed). 1708. *Cuaderno de arquitectura*. Ms. 9114, Biblioteca Nacional de España (BNE).

MARTÍNEZ DE ARANDA, GINÉS (attributed). ca. 1600. *Cerramientos y trazas de montea*, Library of the Servicio Histórico del Ejército, Madrid. (Facsimile ed. Madrid: CEHOPU, 1986.)

VANDELVIRA, Alonso de (attributed). ca. 1575. *Libro de trazas de cortes de Piedras*. Ms. R31, Library of the Escuela Técnica Superior de Arquitectura de Madrid (ETSAM).

### Printed works

BARBÉ-COQUELIN DE LISLE, G. 1977. *Tratado de Arquitectura de Alonso de Vandelvira*. Albacete: Caja de Ahorros de Albacete.

———. 2008. Creatividad y sumisión al poder de la Iglesia: Juan de Portor y Castro y sus Cuadernos de Arquitectura manuscrito, un testimonio ejemplar (1708-1719). Pp. 63-77 in *Arte, poder y sociedad en la España de los siglos XV a XX*, Miguel Cabañas Bravo, et al., eds. Madrid: Ediciones Consejo Superior de Investigaciones Científic.

CALVO LÓPEZ, J. 1999. Cerramientos y Trazas de Montea de Ginés Martínez de Aranda. Ph.D. thesis, Universidad Politécnica de Madrid.

———. 2002. Superficies regladas desarrollables y alabeadas en los manuscritos españoles de cantería. IX Congreso Internacional Expresión Gráfica Arquitectónica, La Coruña.

DE L'ORME, Philibert. 1567. *Premier tome de l'Architecture*. Paris: Federic Morel.

FREZIER, Amédée-François. 1737-1739. *La théorie et la pratique de la coupe des pierres et des bois … ou traité de stéréotomie a l'usage de l'architecture*. Strasbourg-Paris: Jean Daniel Doulsseker-L. H. Guerin.

PALACIOS, J. C. 1990. *Trazas y Cortes de Cantería en el Renacimiento Español*. Madrid: Ministerio de Cultura.

RABASA DÍAZ, E. 2000. *Forma y Construcción en Piedra. De la cantería medieval a la estereotomía del siglo XIX*. Madrid: Akal.

RABASA DÍAZ, E. and J. M. GENTIL BALDRICH. 1996. Sobre la geometría descriptiva y su difusión en España. Pp. 55-93 in Gaspard Monge, *Geometría Descriptiva* (facsímile of 1803 Spanish ed.). Madrid: Colegio de Ingenieros de Caminos, Canales y Puertos.

SAN NICOLÁS, F. Lorenzo de. 1639-1664. *Arte y uso de arquitectura*. Madrid (Facsímile ed., Madrid: Albatros Ediciones, 1989).

TAÍN GUZMÁN, M. 1998. *Domingo de Andrade, maestro de obras de la catedral de Santiago (1639-1712)*. La Coruña, Ediciones do Castro.

TOSCA, Thomás Vicente. 1707-1715. *Compendio Matemático*. 9 vols. Valencia.

———. 1727. *Tratado XV de la montea y cortes de cantería*. (Facsímile ed., Valencia: Librería París-Valencia, 1998.)

## About the author

Rocío Carvajal Alcaide is an architect. She graduated from the ETS of Architecture of Madrid (Universidad Politécnica) and has had a professional practice since 1997. She is a lecturer in the area of architectonic graphical expression at the Centro de Enseñanza Superior CEU-Arquitectura (Universidad Politécnica de Madrid), and at the Escuela Politécnica Superior de la Universidad CEU-San Pablo de Madrid since 2001. She lectures in the course "The art of stone: theory and practice of masonry" in the II Summer School, Universidad CEU San Pablo. Her research involves the history of descriptive geometry, construction, stereotomy and stonecutting craft and techniques. At present, she is developing her doctoral thesis concerning the architecture notebook of Juan de Portor y Castro (BN Mss 9114). She is a participant in the research project "Stonecutting in the Mediterranean and Atlantic areas. Analysis of built examples", financed by the Spanish Ministerio de Educación y Ciencia, directed by Enrique Rabasa Díaz (2010-2012).

Miguel Alonso Rodríguez

Escuela Técnica Superior de
Arquitectura
Universidad Politécnica de Madrid
Avda. Juan de Herrera 4
28040 Madrid, SPAIN
miguel.alonso@upm.es

Elena Pliego de Andrés

Escuela Técnica Superior de
Arquitectura
Universidad Politécnica de Madrid
Avda. Juan de Herrera 4
28040 Madrid, SPAIN
elena.pliego@upm.es

Alberto Sanjurjo Álvarez*
*Corresponding author

Escuela Politécnica Superior
Universidad CEU San
PabloUrbanización Montepríncipe s/n
28668 Boadilla del Monte
Madrid, SPAIN
asanjurjo@struere.es

Keywords: stereotomy, stonecutting
treatises, axonometry, voussoirs,
drawing techniques, representation

Research

## Graphical Tools for an Epistemological Shift. The Contribution of Protoaxonometrical Drawing to the Development of Stonecutting Treatises

Presented at Nexus 2010: Relationships Between Architecture and Mathematics, Porto, 13-15 June 2010.

**Abstract.** In this paper we analyze the graphical methods employed in the stonecutting treatises from that of De l'Orme, published during the sixteenth century, to De La Rue's *Traité de la coupe des pierres*, published in 1728. We will deal with the graphical means that made possible an epistemological shift between treatises of the sixteenth century, written for practising stonecutting masons in a period where the subject was covered by a veil of secrecy, which in some specific cases included a very intuitive form of representation, "protoaxonometries", and eighteenth-century treatises, which show a clear scientific and pedagogical approach to the geometrical problems posed by stonecutting.

## Introduction

In 1728 the French architect Jean Baptiste De La Rue published the first edition of his *Traité de la Coupe des Pierres*. Focused on the stonecutting technique or stereotomy, this treatise had a great influence on the subsequent development of this science and its didactic character, basically in France.

Although its author was not given much consideration by the French architectural historiography, his treatise constituted the basis for the pedagogical discourse of Gaspar Monge, as certain authors report, and it was reprinted twice during the eighteenth and nineteenth centuries (1764 and 1858).

What was the reason for the success of this treatise? Certainly, the care with which it was prepared and the high quality of its graphics, which found no parallel in previous treatises, helped its diffusion, but its intention to make this science more accessible to all professionals, and not only to the masons who were specialized in the stonecutting techniques, constituted the real value of De La Rue's treatise.

Among the graphical tools employed to achieve this object, we will point out the systematic use of axonometrical perspectives to depict the construction elements developed and their component parts. It is here that are found the significant contribution of De La Rue and the main reason for the success of his treatise.

## A code for experts

*Le secret de l'architecture*, as Jousse entitled his treatise [1642], or *el perpetuo silencio*, mentioned by Martínez de Aranda (c.1600), are expressions that show how close in spirit the first stonecutting treatises and manuscripts, such as those of De l'Orme, Vandelvira, Martínez de Aranda, Jousse, and others, were to the medieval tradition. As a consequence of this professional secrecy, mentioned by Philibert De l'Orme [1567: 50], who proclaimed the advantages of knowledge transmission through direct practice, the graphical methods used in this period are based on a very abstract mode of representation, difficult for anyone to understand, whose principal object is to obtain relevant geometrical information for construction.

These geometric constructions of the first stonecutting treatises are based on orthographic projections: plan and elevation. Auxiliary constructions, such as changes of projection plane and rotations, are other, complementary graphic devices. According to several authors [Sakarovitch 1998; Rabasa 1999; Calvo López 1999], these procedures can be considered as the germ of the science of descriptive geometry.

However, this method doesn't appear for the first time in the technical treatises of the sixteenth century in France and Spain. Rather, it derives from the Gothic method used for the construction of ribbed vaults, based on the tracing on the ground of the full-size figure of the plan of the vault and the pattern of its ribs, obtained by rabatting the elevation onto a horizontal plane, so that one could easily relate the levels of the different arches.

These drawings, as described by Rodrigo Gil (BNE Ms. 8884, c. 1540), were traced on working platforms of the scaffolding situated under the vault to be constructed, wooden works that were removed after the construction, so that very few examples of this kind of tracings have survived. Fortunately, we have two Spanish drawings that seem to be sketches similar to those used in the tracing of medieval cross-ribbed vaults: one of them in the manuscript of Hernán Ruiz, and the other one from the vault on the choir of Priego de Córdoba (although this was made for the measurement of the construction works) [Rabasa 1999b: 197; Rabasa 1999a: 130-131]. In any case, these full-size schemes, where the plan and the levels of the different arches can be related on the same drawing, were very abstract and simple.

While some authors consider this Gothic method to be the immediate predecessor of the graphical procedures employed during the Renaissance for the construction of stone vaults, they didn't still constitute a real language, since it was not possible to deduce from them a general and universal method to be understood by everyone [Sakarovitch 1998; Pérez Gómez 1983].

During the eighteenth century, the medieval "veil of secrecy" gave way to illustrated science, and the pedagogical interest of stereotomy became relevant. The treatises were not focused solely on practical stonemasonry, as in the previous period; a new social group, more refined and cultivated, such as mathematicians, engineers and architects, also showed concern for these geometrical problems.

In most cases, the general view of the architectural element was represented through double orthogonal projection, and the details, such as the shape of the voussoirs, was developed by means of some kind of axonometric view, since this provided a visual representation and not only facilitated understanding of the volume of the piece, but also the process of stone carving, especially in the case of the squaring method. These kinds of graphical resource will become fundamental tools for the conceptual changes of the new treatises on stonecutting.

De La Rue's *Traité de la coupe des pierres* [1728] is the first one in which the general use of very accurate axonometries is found. Each architectural element is graphically developed with orthogonal projections and a general view in perspective, a mode of representation that was already found in Desargues [1640]. To explain the process of stone carving De La Rue most often uses a military axonometry, a cavalier perspective or a pseudo-Egyptian perspective. But this resource had been already explored in the sixteenth-century treatises, which in some specific cases included a very intuitive form of this method of representation, which we will call protoaxonometries.

In this paper, we analyze the graphical methods employed in the stonecutting treatises from the sixteenth to the eighteenth centuries in order to obtain clues about the influence of the axonometry in the epistemological shift that took place in the science of stonecutting.

## Axonometry and parallel projection

The term axonometry was used for the first time in 1852 in Meyer and Meyer's *Lehrbuch der axonometrischen Projectionslehre* [1852] to describe the drawing method based on parallel projection, where the projectors are perpendicular to the picture plane, to distinguish them from the oblique projection, where the projectors are oblique to the picture plane, which is the case of cavalier or military perspective, of common use in the French tradition [Loria 1921; Bryon 2008].

The drawings that we analyze here were made before the eighteenth century and are therefore previous to the scientific systematization of the method. They were made in a very intuitive way, and were probably not understood in their day as projections at all, but as situating the different points to represent in an axis system or simply by maintaining one plane undistorted (the elevation in the case of the cavalier perspective, or the plan in the military perspective) and practising a kind of extrusion of this face.

For the drawings of the first stonecutting treatises mentioned here, we consider the term protoaxonometries more appropriate, or, in some cases, parallel projections. The terms cavalier and military perspectives, as used traditionally in France, will also be employed to describe the parallel projections where the picture plane is parallel to one of the faces of the represented object.

In the case of the drawings of De La Rue's treatise, it is clear that, because of their graphical development and their quality, we can already speak of axonometries.

## The use of parallel projection in the history of the drawing techniques

We can find multiple examples of drawings using oblique projection as early as Antiquity, where we find them in Greek vase paintings and in the Roman frescoes of Pompei.

It is, however, in China where this technique was introduced as the main method for depicting three-dimensional space and where the construction drawings developed its early use, since central perspective was not known in China until the seventeenth century, when it was introduced by the Jesuits.

Long before Leonardo introduced the aerial view or bird's-eye perspective in Western painting, the Chinese used it with a great ability. The painter and writer Huo His (tenth century) was the first one to systematize the three-dimensional representation in landscape views [Scolari 1984].

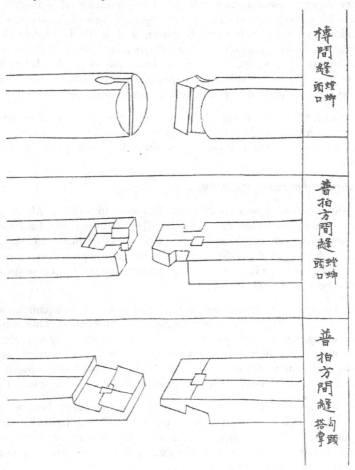

Fig. 1. Joining in tie-beams, *Ying Tsao Fa Shih* (ca. 1103), ch. 30, p. 17b

The Chinese *Ying Tsao Fa Shih* construction treatise of 1103 already contained excellent construction drawings. We can almost speak of working drawings in a modern sense (fig. 1), perhaps for the first time in any civilization [Needham 1971], unparalleled in the West, developed by means of the parallel projection technique, a system in which lines that were parallel in fact remained so in the drawing, in a way that we can say is very similar to that used today by architects and engineers for mechanical or structural "working drawings." This way of representation is indicative of an attitude towards Nature that is at once humbler and more social than that of Western man [Needham 1971].

European culture had to wait until Leonardo's exploded views (fig. 2) to find technical drawings using some kind of parallel view, which is a type of drawing very common later on.

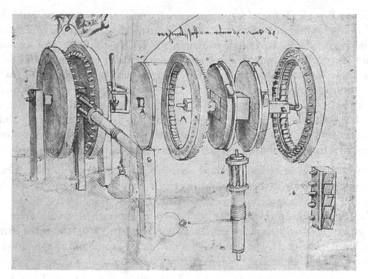

Fig. 2. Leonardo da Vinci, *Codex Atlanticus*, detail, fol. 30v/8v-b

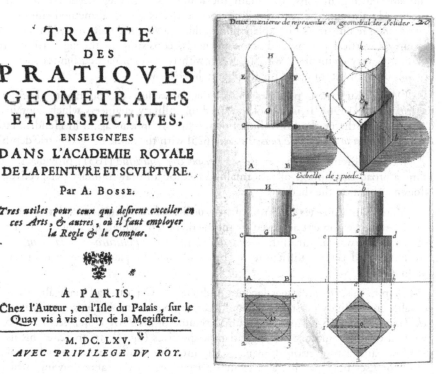

Fig. 3. Abraham Bosse, *Traité des pratiques geometrales et perspectives* [1665],
Title page and pl. 20

Abraham Bosse's *Traite des pratiques geometrales et perspectives*, published in Paris in 1665 and organised in two parts, can be understood as a textbook or a notebook of his lessons in the Academie Royale de la Peinture et de la Sculpture. The first part, focused on theory and with few graphical references, is addressed to painters and designers, and contains some notes about the military perspective. Bosse mentions two forms of representation: *geometral* and *perspectif.* The term *geometral,* which is used nowadays in France to designate the orthographic projections, can be understood as a synonym of *petit pied* or scale, since he uses it to refer to the possibility of measuring and, therefore, to a type of drawing in which in some way certain geometrical attributes of the object remain undistorted, as opposed to the term *perspectif,* in which they appear modified.

The so-called geometrical method refers to the representation by means of both orthographic projections (the plan and the elevation), and by means of the development or unfolding of the faces on a horizontal plane, as we can see in the military perspective of a cube (fig. 3). Bosse insists that both forms of representation are *deux especes d'un mesme genre, et non pas deux genres divers* (two spaces of the same kind, not two different kinds) [1665: 3]. He also explains the method to obtain the shadows to allow an increasingly solid appearance.

In the practical part of the treatise, profusely illustrated with graphical details, he uses military perspective to explain the development of a solid by rabbating its faces to a horizontal plane, a method that, as he indicates, was used by some professionals, such as the stonecutting masons, to obtain the templates that they needed to cut the stone blocks. He also maintains that military perspective is the usual method employed by the engineers. He illustrates his explanation of this drawing method with figures of prisms or groups of vertical prisms and some pyramids, consisting in tracing verticals from the plan, in a very elementary way, to go on with more complicated figures such as sloped prisms or complex solids, to finally obtain the shadows.

One documented predecessor of this procedure is Buonaiuto Lorini's *Delle fortificationi* [1596]. It is a book destined for military engineers which contains a brief exposition of the method of military perspective. It appears again in Hendrik Hondius's *Instruction en la science de perspective* [1625] with the name of "geometric depth".

Jean Du Breuil, in his treatise *La perspective pratique* [1651] provides for the first time a broad description of a normalized method to draw military perspectives, but he does not mention the shadows.

More than a quarter of Bosse's text is dedicated to military perspective, showing the significant development achieved by this mode of representation, which had an empirical character and was taught in the *Academie Royale de la Peinture et de la Sculpture,* but, as we know, had certain difficulties in being adopted by painters or even in the field of architecture.

Despite this, the architect Jacques Androuet du Cerceau had made already use of the cavalier and military perspective about a century before. In *Les plus excellents bastiments de France* [1576-1579] and in the third volume of his *Livre d'architecture* [1559-1582] we find cavalier, military and Egyptian perspectives (fig. 4). In any case, his use of this kind of parallel projection drawing is very intuitive, and in many cases he makes some errors or uses both parallel and central perspective on the same drawing, which shows that he had not entirely mastered this method of representation.

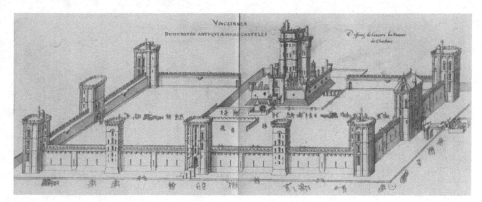

Fig. 4. Jacques Androuet Du Cerceau, Vincennes, from *Les plus excellents bastiments de France* [1576-1579], Bk. I, pl. XXVI

In *Des fortifications et artifices architecture et perspective* from Jacques Perret [1601] we find again the use of military perspective but in this case applied to the field of military architecture, since it deals with polygonal bulwarks that are represented in the plates by tracing verticals from the plan, a procedure which can be related to the method described later by Bosse [Alonso 1991].

### First predecessors in the use of protoaxonometrical drawings in stonecutting treatises

In the sixteenth-century treatises appear some drawings, which we will call protoaxonometries, made by the extrusion of the two-dimensional image of the elevation or the plan, resulting in an oblique parallel projection. In these representations, the plane of the object that is parallel to the picture plane remains undistorted, making it a form of representation particularly suitable for illustrating stonecutting techniques. In these first examples that we will analyze, the method employed to trace them is very intuitive and practical, rather than geometric or mathematic, since we will have to wait until the following century [Bosse 1665] to find the first attempts to systematize the technique of drawing parallel projections.

Among the first construction drawings made as parallel projections, or protoaxonometries, are those found in Vandelvira's manuscript of 1575 ca. [Ms. R31] to explain the cutting of the steps in spiral staircases. Fig. 5 (left), which corresponds to *declaración del caracol exento*, shows two steps seen from below, but rotated to show the intrados surface, in a protoaxonometrical drawing of the horizontal picture plane. The two steps, while not connected with vertical lines, are situated so as to show how they join together; this could thus be considered an "assembly drawing" in a modern sense.

This graphical resource is frequently found in construction treatises, especially in wood joint drawings, such as those of the Chinese treatise *Ying Tsao Fa Shih* from the twelfth century. At the beginning of the nineteenth century, Rondelet universalized it in his *Traité théorique et pratique de l'art de bâtir* [1802], one of the first books on general construction. In the stonecutting treatises it will be frequently used beginning with Frezier [1737-1739], but the first to make use of it was De La Rue [1728].

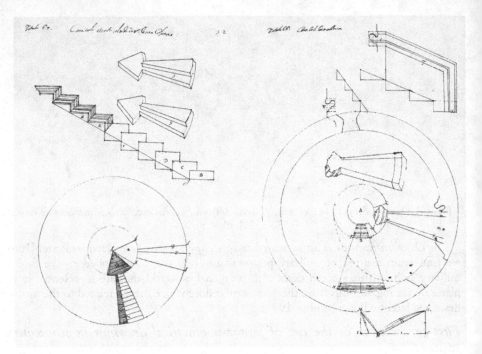

Fig. 5. Ms. R31, known as *Libro de trazas de cortes de Piedras*, attributed to Alonso de Vandelvira. Left, *Declaración del caracol exento*, fol. 51v; right, *Declaración del caracol de Mallorca*, fol. 49v

It is significant that, to represent the intrados surface of the steps, instead of constructing a worm's eye perspective, a technique that he probably did not know, Vandelvira draws a bird's-eye perspective, which forces him to turn the piece. This recourse will appear again in other authors, such as De La Rue, as we will see later.

In Vandelvira's drawing dedicated to the *caracol de Mallorca* (fig. 5, right) we find another step represented in a protoaxonometrical view of the horizontal picture plane, where the dimensions of the piece were taken directly from the plan. In this case, Vandelvira makes a drawing that we would now call *wireframe*, and indicates the part of the step that fits into the wall. It is the initial volume previous to the carving of the helicoid of the soffit surface.

However, Vandelvira, who wrote his treatise between 1575 and 1580, was not the first to employ the protoaxonometrical drawing for stonecutting tracings. These types of protoaxonometries are also found in three plates in the first stonecutting treatise published in 1567 by Philibert De l'Orme: the *porte biaise, the porte sur l'angle*, and *the porte dans une paroi de plan courbe* (fig. 6). In all three cases, he develops the detail of one or two voussoirs in parallel projection to show its volume, in order to make the object understandable as a three-dimensional entity in space. These are cavalier projections that exhibit an orthographic view or elevation of the front size, and therefore this face is represented in true shape.

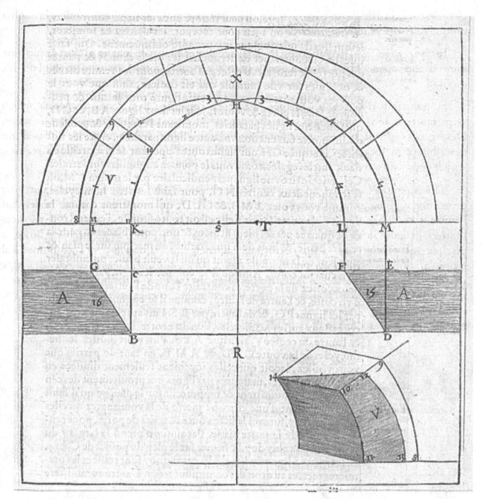

Fig. 6. Philibert De l'Orme, *Premier tome de l'Architecture* [1567], Bk. III, fol. 69r, *porte biaise*

Something similar is found in Martínez de Aranda's manuscript [1600 ca.]. At the beginning of the second part, focused on arches and splays, he uses parallel projection to explain the squaring method in four cavalier perspectives (fig. 8, left), but in this case seen from below. Two of them show the initial prismatic volume on the front face of which is drawn the face template of a voussoir as a sector of circular crown which, after being squared straight, results as a voussoir of an arch, and after being worked curved results as a dome. In the first case, the edges are straight lines; in the second one they are arches. Except for their inclusion at the beginning of the second part, this kind of drawing does not appear anywhere else in the treatise. These drawings are interesting since they constitute the earliest examples of worm's eye views found in stonecutting treatises.

## The didactical approach: Desargues – Bosse – De La Rue

In 1640 Girard Desargues published a short work entitled *Brouillon Project d'exemple d'une manière universelle touchant la pratique du trait à preuves pour la coupe des pierres…* in which he tries to solve the particular problems on stereotomy through a unique rule: the *manière universelle*.

In four pages and five figures he analyses a sloped vault that is at the same time skewed and opened to a sloped wall. This is the only model he uses to explain his method.

Stonecutting treatises, up to then, had been organized as a series of models which were more or less hierarchical, but did not establish a general method. In this sense, Desargues's approach was in some way didactical and epistemological, focused on describing a methodology [Bessot 1994: 297], clearly in contrast to the previous treatises, which described stonecutting work from a geometrical point of view.

This method contains an important graphical development. Desargues starts by presenting a global image through a central perspective of the architectonical model that he wants to study. Up to then, stonecutting treatises had not been devoted to these matters [Sakarovitch 1994: 351; Sakarovitch 1998: 174]. In 1643 Abraham Bosse, a disciple of Desargues who dedicated himself to problems of axonometrical representation, published *La pratique du trait a preuves de M. Desargues pour la coupe des pierres en l'architecture*, where he tries to apply the method of his teacher to a great number of particular cases: sloped vaults, skewed arches, corner arches, squinches, splays, etc. Bosse carefully follows the indications of Desargues, not only applying his method but also in his presentation of the different models to study.

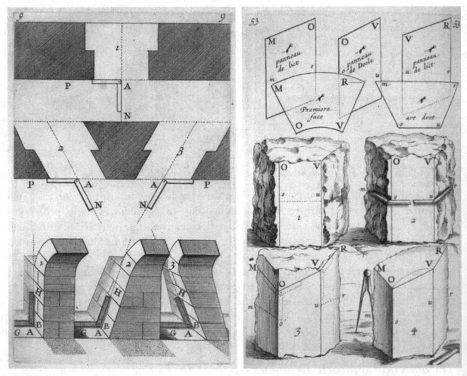

Fig. 7. Abraham Bosse, *La pratique du trait à preuves de M. Desargues,…* [1643], Pl. 53

He makes many global perspectives of the construction elements that he analyses, some of them in cavalier perspective (fig. 7 left), but he also shows a new mode of representation that will be used by Derand [1643] and further developed by De La Rue. We are referring to the parallel projection views that show the stonecarving method showing the tools used on the process, as we can see in plate 53 (fig. 7 right). While De La Rue will not accept Desargues's thesis, he will adopt and develop this pedagogical mode of representation in his treatise.

It is obvious that De La Rue knew Bosse's work regarding the stonecarving process. He mentions it in the preface of his treatise and we can also presume that he knew Bosse's treatise about geometrical representation [1665], since we have found important common points, such as the employment of the developed faces rabatted to a horizontal plane of a prism to show the application of the face or the bed templates of a voussoir.

De La Rue will follow the scheme of presenting first a general perspective or axonometry of the model and the plan or the elevation and, to facilitate comprehension of the stonecarving process, some views with the detail of the sometimes complex process. This scheme, as we can see, is based on the method proposed by Desargues and further developed by Bosse.

However, based on a new didactical spirit, De La Rue will develop these ideas much further, and therefore he contributed with them to the important epistemological shift that took place in stereotomy.

## Graphical connections between Derand and De La Rue

In 1643 François Derand published a treatise devoted to stereotomy, *L'architecture des voutes ou l'art des traits et coupe des voutes*, probably the consequence of his teaching activity at the Collège in La Flèche; Jousse probably borrowed Derand's ideas to a great extent in *Le secret d'architecture*, although this was published one year before [Pérouse de Montclos 1982]. Both treatises deal with the art of stonecutting in relation with a collection of examples, but with a practical approach, very different from the scientific approach adopted by Desargues in his *Brouillon projet d'exemples d'une manière universelle*.

In Derand's treatise the use of parallel projection is not very common, but it does contain some drawings that can be considered precursors of the work developed later by De La Rue, such as, for example, those used to explain the squaring method in the voussoirs of cavalier and military perspectives, where depth is indicated by an oblique extrusion, from the elevation or from the plan. We also find in these drawings new and very expressive graphical resources: Derand shows the initial volume of the stone piece to cut the voussoir and indicates the parts to eliminate to obtain the final shape. We will find this new code, much more elaborated, also in De La Rue's treatise [1728].

In the first drawing where we find the use of the cavalier perspective, Derand makes use of it to explain in a general way the squaring method in the cutting of the voussoirs (fig. 8 right). In this drawing he traces a prism, in a way that we would call now wireframe, on the front face of which he draws the template, tracing parallel lines from its vertex to the edges of the prism, to obtain an image where the parts to eliminate from the solid are shown, in a very similar way to that developed previously by Martínez de Aranda (fig. 8 left).

Again, in plate IX dedicated to the *biaise par teste par équarissement*, he will resort to parallel projection to show the volume of the three voussoirs that are part of the arch. In this case he uses Egyptian perspective, and we can recognize the elevation, and therefore the face template and the plan both in true shape.

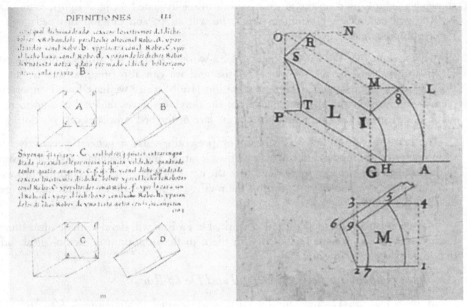

Fig. 8. Left, Martínez de Aranda, *Cerramientos y trazas de montea* [1600 ca.]; right, Derand, *L'Architecture des voûtes* [1643]

Although in the first voussoir he shows the whole volume of the initial solid as a wireframe prism, in the second one he shows only a little part of it, and in the last one he shows only the end shape of the voussoir. Each of the three voussoirs appears in its real orientation. We can affirm that the use of this kind of Egyptian perspective is not at all usual in these treatises, although we will find occasionally some of them in De La Rue.

We find the use of parallel perspectives again in the plate dedicated to the *biaise passé*. The drawing corresponds to the *Corne de boeuf* and the method he uses to arrive at the parallel perspective is obvious: from the elevation he takes the projection of the first voussoir, the front but also the back face, and traces from its vertex lines parallel to the depth, in a kind of extrusion as in a cavalier perspective, to define the vertex of the first voussoir of the arch, which leads to an image that shows at the same time the voussoir and of the solid volume from which it comes. He uses the same code again to indicate the parts to eliminate from this initial solid by a dashed line, as we saw in the previous plate and as we will see, more elaborated, in De La Rue.

In a subsequent plate, the one dedicated to the *Voute d'arestes barlongue par equarrisement*, Derand uses a military perspective from below for the two first voussoirs on the groin (fig. 9). In fact, what he shows is the whole initial solid in order to trace on its faces the silhouette that will allow it to perform as the groin of the vault. It is significant that in one of these perspectives are shown the *x*-, *y*- and *z*-axes, with the *x*-axis being the vertical one.

This treatise, in which drawings in parallel projection appear only very occasionally, will have a decisive influence, partly because of the use that Derand makes of it, but also because of the graphical resources that we can see here, still in a very primitive way.

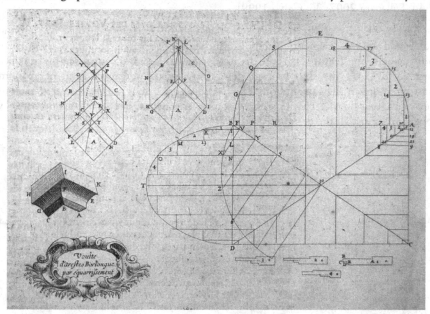

Fig. 9. Derand, *L'Architecture des voûtes* [1643], *Voute d'aresies barlongue par equarrisement*

## Axonometrical drawing in De La Rue's Traité de la coupe des pierres

De La Rue divided his treatise into five chapters that were accompanied by a preface and a *petit traité de stereotomie*. In the preface, De La Rue states his intentions. After briefly discussing the state of the art, surveying the works that had been published up to that point, he arrives at the main objective of his publication: to make the treatise, and therefore, the science it intends to communicate, instructive and comprehensible. In order to obtain this objective, the treatise was carefully laid out, and each model studied is accompanied by an elevation, the design of the centering and an axonometric projection of the voussoirs in the different steps of construction.

> *Comme les Livres qui traitent de la Coupe des pierres sont très rares aujourd'huy; & que d'ailleurs, entre ceux qui ont paru jusqu'à present, il n'y en a pas un qui soit assez a 'la portée, soit des Commençans ou des Ouvriers, j'ay crû qu'il ne seroit point inutile de presenter celuy-cy au Public. Je n'ay rien negligé pour le rendre aussi instructif qu'intelligible : & afin d'y parvenir plus sûrement, j'ay accompagné la plus grande partie des Epures, de leur élevation & de leur cintre ; j'y ay joint la représentation de plusieurs de pierres tracées dans des degrés differens, pour suppléer à la foible idée qu'en donne une Epure embarrassée des lignes qui la composent ; ce qu'on n'avoit pas encore fait jusqu'icy* [De La Rue 1728: Preface].

The *petit traité de stéréotomie* included in the book analyses several theoretical aspects that are related to spatial geometry and, in particular, to the plane sections of cones, cylinders and spheres and the development of cones and cylinders. It is in fact the first treatise to include the term "stereotomy" in its title. This term appeared for the first time in the booklet written by Jacques Curabelle against Desargues in 1644 [Curabelle 1644; Fallacara 2007: 36; Calvo 1999].

The main part of the treatise is arranged in five chapters devoted to the study of arches and splayed elements, vaults, squinches, sloped vaults and stairs. In total, sixty-four models are featured in the treatise, whereas Derand's treatise, his predecessor, featured 124. Of the sixty-four models, thirty-six include some kind of axonometric drawing to aid in the explanation. The fourth chapter does not include any axonometric views, whereas in the chapters about vaults, squinches and stairs, axonometric views appear in almost all the designs.

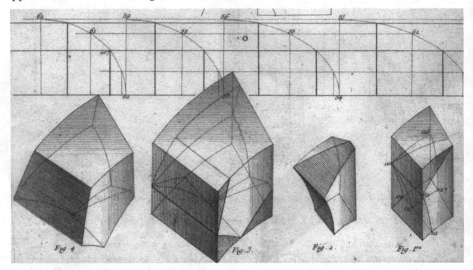

Fig. 10. De La Rue, *Traitè de la Coupe des Pierres* [1728], pl. XXX (detail)

The layout of each chapter also responds to a didactic criterion. De La Rue first explains those construction elements that are easier to carve. When the reader then comes to more complex problems, he already understands the basic techniques necessary for the comprehension of these harder problems. However, this is not an innovation, for this arrangement was already present in Spanish stonework manuscripts of the sixteenth and seventeenth centuries.

De La Rue frequently uses axonometric views to explain the carving process for the different types of voussoir and this becomes a key tool to comprehend the sometimes complex stone cuts (fig. 10).

The technique used De La Rue for drawing parallel projections of the pieces is rather simple. The different vertices of the piece are transferred from the top view or elevation onto the projected view using their space coordinates. This technique can be used to obtain any view, no matter how complex. Coefficients of reduction are not used, nor are projections of coordinate planes. It is a mere translation of points from one drawing to another. Therefore, if we were to apply modern definitions of descriptive geometry, we would not consider De La Rue's representations axonometric views in the strict sense of the term.

According to the type of piece and the face of the piece to be represented in true shape, De La Rue uses various types of axonometric views, all of which are oblique views, in keeping with the French tradition. It is worthwhile recalling that the orthogonal projections described by William Farish in 1822 would be mainly developed in Great Britain and Germany.

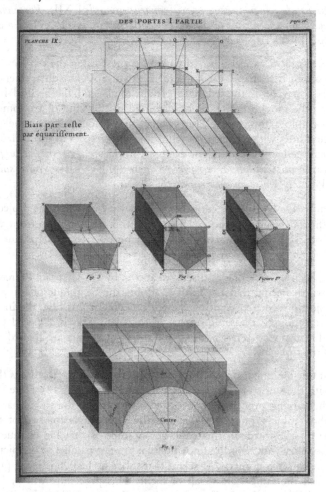

Fig. 11. De La Rue, *Traitè de la Coupe des Pierres* [1728], pl. IX, *biais par teste par équarissement*

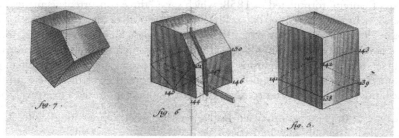

Fig. 12. De La Rue, *Traitè de la Coupe des Pierres* [1728], suite de la planche 65, *escalier suspendu et a repos* (detail)

There are examples in cavalier perspective, military perspective and pseudo-Egyptian perspective. When using cavalier perspective, the front vertical planes are in true shape and the top view is deformed. It is interesting the fact that the projection of the *y*-axis is drawn to the left in almost all cases, the opposite of what is common in this kind of representation. Examples of cavalier perspective can be found, for example, in *Porte droite en tour ronde par panneaux et par equarissement* (pl. XII), *Porte en tour biaise en talut etrachetant une voute sur le noyau par equarissement* (pl. XVI), *La corne de Boeuf* and *Biais Passé* (pl. XVII).

Military perspectives are also broadly used. In these cases, the top view and the planes parallel to it are in true shape. De La Rue uses two types of military perspective. In the first case, the top view is rotated by a certain angle and the vertical lines are drawn vertical; this is the most common technique used in this kind of axonometric views. De La Rue also employs this type of representation without rotating the top view, thus drawing vertical lines with a constant inclination with respect to the horizontal. Examples of military perspectives are found, for example, in *Voute d'Areste barlongue* (pl. XXIV), *Voute spherique ou Cu-deFour en plein Cintre* (pl. XXVII), *Voute sur le noyau* (pl. XXIX), and *Voute d'Areste en tour ronde* (pl. XXX).

Furthermore, De La Rue uses another type of view in which the projections of the *y*-axis and of the *z*-axis are superposed. Thus, the top view and top front view appear on top of each other. These representations could be referred to as pseudo-Egyptian (fig. 11).

They had been used before, as discussed above, by Du Cerceau, Perret, Dubrueil and Bosse. It is worth noting that De La Rue adopts the technique of using these pseudo-Egyptian projections for representing skewed objects. In this case, the inclination of the horizontal projections is the same as that of the plan. This is a special type of military perspective. The top view is not rotated, and the vertical lines are aligned with one of the directions of the top view.

The most outstanding axonometric projection found in the treatise is that of pl. LXV, concerned with the study of *l'escalier suspendu et a repos* (fig. 12). Here De La Rue tries to explain the carving process of one of the voussoirs by drawing what seems to be a dimetric orthographic axonometry: he uses the same angle for two of the axes and a different one for the third. In this case, our opinion is that De La Rue is drawing an oblique axonometry with coordinate planes that are not parallel to the picture plane, that is, a generic oblique axonometry. It seems that, once he had chosen the projection of the axes, he applied the same scale to all of them. This procedure will be broadly used, a century and a half later, by the French engineer Auguste Choisy. It is an application of Pohlke's theorem, formulated in 1858, which says: "three concurrent segments in a plane can always be considered as the cylindrical projection of the three edges of a cube" [Rabasa 1999]. It is obvious that De La Rue did not know this theorem, formulated almost two centuries later, but his procedure for constructing parallel projections is closer to these oblique representations than to the procedure for obtaining dimetric axonometries, such as the English or German axonometries of the nineteenth century. These kinds of oblique representations frequently appear in Frezier's treatise [1738].

A type of drawing mainly used by De La Rue when representing vaults is what we earlier called "assembly drawing" (fig. 13).

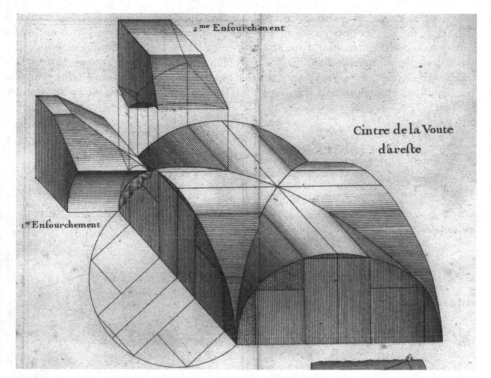

Fig. 13. De La Rue, *Traité de la Coupe des Pierres* [1728], detail, pl. XXIV, *cintre de la voute d'areste barlongue*

As mentioned, this type of drawing has been widely used in engineering and construction. De La Rue usually makes use of this representation together with an axonometric projection of the vault centering. This way he achieves two objectives: on the one hand, the intrados surface of the vault is represented, as it is the same as that of the centering, and on the other hand, the centering is used as a support element, helping to organize the assembly of the different pieces that form the vault.

In this case, and in others that simply describe the cutting process for a voussoir, De La Rue uses bird's-eye views. We cannot say that he did not know how to make a worm's-eye view, since we can find at least three examples of this kind of perspective in the treatise – *l'Arriere Voussure de St. Antoine, en plein cintre* (pl. XX), *Voute spherique ou Cu-de-Four en plein Cintre* (pl. XXVII) and *Trompe en niche rampante* (pl. XLVII) – but there is no doubt that he was more comfortable drawing bird's-eye views. This is the case of the description of the carving process of certain voussoirs, such as those of the plate related to the groin vault (pl. XXIV, fig. 13), where he prefers to rotate the piece while maintaining the same type of projection than to change to a worm's-eye view to see the piece from below. This is very common in the few protoaxonometrical drawings in the stonecutting treatises before De La Rue's. With the exception of the drawing by Martínez de Aranda mentioned before, all previous drawings on stonecutting treatises are bird's-eye views. This fact makes the drawing by Martínez all the more interesting. Not until Frezier and the experts on descriptive geometry in the nineteenth century will worm's-eye view axonometries be used widely in stereotomy treatises.

Another remarkable characteristic of De La Rue's special way of representation is found in military perspectives with the face or bed templates rabatted to a horizontal plane to show them in true shape (fig. 14, right). This kind of drawing responds to a clear pedagogical intention and has its antecedents in the drawings that we have already mentioned from Bosse's *Traite des pratiques geometrales et perspectives* [1665], where he represents a series of prismatic figures with its faces rabatted to a horizontal plane (fig. 14, left).

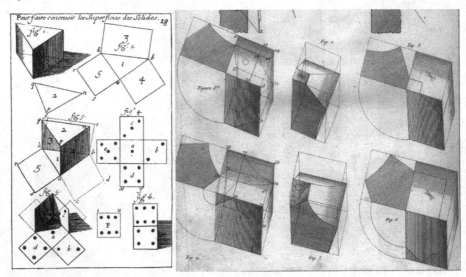

Fig. 14. Left, Bosse, *Traite des pratiques geometrales et perspectives* [1665]; right, La Rue, *Traitè de la Coupe des Pierres* [1728], pl. XXIV, *voute d'areste barlongue*

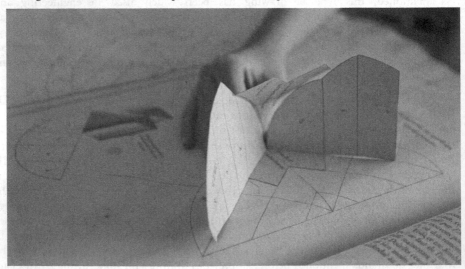

Fig. 15. De La Rue, *Traitè de la Coupe des Pierres* [1728], fold-outs, *Preuve de l'erreur du premier panneau d'enfourchement de la voute spherique fermée en quarré, dévelopé suivant Philibert Delorme, Mathurin Jousse a & le Pere Derand*

Certain graphical devices used by De La Rue are indicative of the same didactical interest, and constitute a genuine innovation in the history of stonecutting treatises. We are referring to the representation of the tools and instruments used in the stone carving process. This device makes it possible for the reader who is not familiar with these techniques to understand the process and to check the application of templets, bevels, straight edges, and so forth.

Another novelty, also indicative of the extraordinary quality of the edition of the treatise, are the fold-outs glued to the plates, used to demonstrate certain theories such as the case of the *Preuve de l'erreur du premier panneau d'enfourchement de la voute spherique fermée en quarré, dévelopé suivant Philibert Delorme, Mathurin Jousse a & le Pere Derand* (fig. 15), where he demonstrates the error commited by the authors named, proven by a development that can be visualized in space.

Apart from its scientific value, without a doubt the most important characteristic of the treatise, that which provides its graphical quality and caused it to be admired, studied and even mutilated so its plates could be sold individually all over Europe, is comprised in the variety of line types (continuous, dashed, dotted), hatching, shading, shadows, etc., which give the drawings a quality and hierarchy of elements never seen before.

## Conclusions

De La Rue, as we have already mentioned, made use of a great number of graphical and spatial devices in his *Traité de la coupe des pierres* to achieve his objective of transmitting the "stonecutting science, one of the most important and useful sciences included in Architecture," and to make it more intelligible not only to the experts but also to those who wish to learn [De La Rue 1728: Preface].

To this end, he makes extensive recourse to the use of parallel projections or axonometries. By then, the old tradition of stonecutting treatises was giving way to the science of stereotomy, and the architects who were devoted to this field gave up their places to engineers. In this sense, Francois Derand's *L'Architecture des Voutes* can be considered the last great stonecutting treatise by an architect.

Traditionally, architect's drawings had been developed through orthographic projections: plan, elevation and section, according to Alberti and the letter to Leo X [Bois 1981]. Architects used to complete their projects with perspectives and also with scale models. Authors of Renaissance stonecutting treatises, also influenced by their medieval heritage, maintained these rules. In contrast, we know about the preference of the engineers for the axonometrical drawing, as Bosse and Dubreuil, among others, maintain.

At this moment of transition in which knowledge about the science of stonecutting was universalized and refined, axonometry plays a fundamental role at the hands of military engineers such as Frezier and, later, by the engineers of the École Polytechnique with Jules de La Gournerie.

Jean Baptiste De La Rue, architect and member of the Académie Royal d'Architecture was also dedicated professionally to the construction of bridges and the invention of machines. So we can affirm that he was an architect-engineer who was key for the epistemological shift which took place during the eighteenth century in France. It is in this sense that his *Traité de la Coupe des Pierres* can be considered the first great treatise on stereotomy by an engineer.

His treatise explored, with great success, the new graphical methods developed by mathematicians and geometers for the study of stereotomy, then further developed by engineers, before its demise as a practical science.

## References

### Manuscripts

DE VANDELVIRA, Alonso (attributed). 1575 ca. *Libro de trazas de cortes de Piedras*. Ms. R31, Library of the Escuela Técnica Superior de Arquitectura de Madrid (ETSAM).

MARTÍNEZ DE ARANDA, Ginés (attributed). 1600 ca. *Cerramientos y trazas de montea*, Library of the Servicio Histórico del Ejército, Madrid. (Facsimile ed. Madrid: CEHOPU, 1986.)

GIL DE HONTAÑÓN, Rodrigo (attributed in part). 1540 ca. Included in Simón García, *Compendio de Arquitectura y simetría de los templos*, 1681. Ms. 8.884, Biblioteca Nacional de España, Madrid.

### Printed works

ALONSO RODRIGUEZ, Miguel. 1991. La axonometría o el espejismo científico de la realidad, práctica y regla como técnica descriptiva. Ph.D. thesis, Universidad Politécnica de Madrid.

ANDROUET DU CERCEAU, Jacques. 1576-1579. *Les plus excellents bastiments de France*. 2 vols. Paris. Rpt. D. Thomson, ed. Paris: Sand & Conti, 1988.

———. 1559, 1561, 1582. *Livre d'architecture, contenant diverses ordonnances de plants et élévations de bastiments*. Paris: Benoît Prévôt.

BESSOT, Didier. 1994. Les aspects épistémologiques de la pensée didactique de Desargues: l'usage des exemples génériques. Pp. 295-312 in *Desargues en son temps*. Paris: Librairie Scientifique A. Blanchard.

BOIS, Y. 1981. Metamorphosis of Axonometry. *Daidalos* 1: 41-58.

BOSSE, Abraham. 1643. *La pratique du trait à preuves de M. Desargues,... pour la coupe des pierres en l'architecture*, par A. Bosse. Paris: imp. De P. Des-Hayes.

———. 1664. *Traité des manières de dessiner les ordres de l'architecture antique en toutes leurs parties: avec plusiers belles particularitez qui n'ont point paru jusques à present, touchant les bastiments de marque*, Paris.

———. 1665. *Traite des pratiques geometrales et perspectives, enseignees dans l'Academie Royale de la Peinture et de la Sculpture*. Paris.

BRYON, Hilary. 2008. Revolutions in space: parallel projections in the early modern era. In *Architectural Research Quarterly* 12, 3-4: pp. 337-346.

———. 2009. Measuring the qualities of Choisy's oblique and axonometric projections. Pp. 31-61 in *Auguste Choisy (1841-1909). L'architecture et l'art de bâtir*. Actes del Simposio Internacional celebrado en Madrid, 19-20 de noviembre de 2009. Javier Giron and Santiago Huerta, eds. Madrid: Instituto Juan de Herrera.

CALVO LÓPEZ, José. 1999. Cerramientos y Trazas de Montea de Ginés Martínez de Aranda. Ph.D. thesis, Universidad Politécnica de Madrid.

———. 2009. La literatura de la cantería: una visión sintética. Pp. 101-156 in *El arte de la piedra. Teoría y práctica de la cantería*. Madrid: CEU Ediciones.

CURABELLE, Jacques. 1644. *Examen des ouvres du Sieur Desargues*. Paris: M. & I. Henault.

DE LA RUE, Jean Baptiste. 1728. *Traitè de la Coupe des Pierres*. Paris: Imprimirie Royale.

DE L'ORME, Philibert. 1567. *Premier tome de l'Architecture*. Paris: Federic Morel.

DERAND, François. 1643. *L'Architecture des voûtes*. Paris: Sébastien Cramoisy.

DESARGUES, Girard. 1640. *Brouillon project d'exemples d'une manière universelle du sieur G.D.L. touchant la practique du trait apreuve pour la coupe des pierres en architecture; et de l'esclaircissement d'vne manière de réduire au petit pied en perspectiue comme en géometral et de tracer tous quadrans plats d'heures égales au soleil*. Paris: Melchoir Tavernier.

DUBREUIL, Jacques, 1642. *La perspective pratique nécessaire à tous peintres, graveurs, architectes, sculpteurs, brodeurs, tapissiers et autres se servant du dessin, par un parisien religieux de la Compagnie de Jésus*. Paris. (2nd edition, vol I, 1651, Paris.)

FALLACARA, Giuseppe. 2007. *Verso una progettazione stereotomica. Nozioni di stereotomia, stereotomia digitale e trasformazioni topologiche: Ragionamienti intorno alla construzione della forma.* Roma: Aracne.

FARISH, William. 1822. On Isometrical Perspective. In: *Transactions of the Cambridge Philosophical Society* 1: 1-19. Cambridge.

FRÉZIER, Amadée-François. 1737-1739. *La theorie et la practique de la coupe des pierres et de bois pour la construction des voutes... ou traité de stéréotomie à l'usage de l'Architecture.* 3 vols. Estraburgo and Paris. (2nd ed. 1754-1769.)

GUILLAUME, J. 1988. Philibert Delorme: un traité différent. Pp. 347-354 in *Les Traités d'Architecture de la Renaissance.* París: Picard.

HONDIUS, Hendrik. 1625. *Instruction en la science de perspective.* Den Haag.

JOUSSE, Mathurin. 1642. *Le secret d'architecture découvrant fidèlement les traits géométriques, couppes et dérobemens nécessaires dans les bastiments.* La Flèche: G. Griveau.

LORIA, Gino. 1921. *Storia della geometria descrittiva dalle origini sino ai giorni nostri,* Milan: Ulrico Hoepli.

LORINI, Benvenuto. 1596. *Delle fortification.* Venice.

MEYER, C. Th. and M. H. MEYER. 1852. *Lehrbuch der Axonometrie oder der gesammten isometrischen, monodimetrischen und anisometrischen Projectionslehre.* Leipzig: Verlag von Otto August Schule. (2nd ed., 1855, *Lehrbuch der axonometrischen Projectionslehre.*)

MILLIET-DECHALES, P. Claude François. 1674. *Cursus seu mundus mathematicus.* Lyon: Anisson.

NEEDHAM, Joseph. 1971. *Science and Civilisation in China.* Vol. 4, part III. Cambridge University Press.

PÉREZ-GÓMEZ, Alberto. 1983. *Architecture and the Crisis of Modern Science.* Cambridge, MA: MIT Press.

PÉROUSE DE MONTCLOS, Jean Marie. 1982. *L'architecture a la francaise. Du milieu du XV a la fin du XVIII siecle.* Paris: Picard.

RABASA DÍAZ, Enrique. 1999. Auguste Choisy: vida y obras. Pp. XI-XXVIII in *El arte de construir en Roma.* Madrid: Instituto Juan de Herrera.

———. 2000a. *Forma y Construcción en Piedra. De la cantería medieval a la estereotomía del siglo XIX.* Madrid: Akal.

———. 2000b. El corte de piedras en el proceso de construcción de las bóvedas góticas y renacentistas. In: *Historia de las técnicas constructivas en España.* Madrid: FCC.

———. 2009. Soluciones innecesariamente complicadas en la estereotomía clásica. Pp. 51-69 in *El arte de la piedra. Teoría y práctica de la cantería.* Madrid: CEU Ediciones.

RONDELET, Jean Baptiste. 1804. *Traité théorique et pratique de l'art de bâtir.* 2 vols. Paris: Chez l'auteur, enclos du Panthéon.

SAKAROVITCH, Joël. 1994. Le fascicule de stéréotomie, entre savoir et metiers, la fonction de l'architecte. Pp. 347-362 in *Desargues en son temps.* Jean Dhombres and Joël Sakarovitch, eds. Paris: Librairie Scientifique A. Blanchard.

———. 1998. *Épures d'architecture. De la coupe des pierres à la géométrie descriptive. XVI^e-XIX^e siècles.* Basel: Birkhäuser.

SCOLARI, Massimo. 1984. La prospectiva gesuita in Cina. *Casabella* **504**: 48-51.

———. 1984. Elementi per una storia dell'assonometria. *Casabella* **500**: 42-59.

———. 1985. Elements for a History of Axonometry. *Architectural Design* **55**: 5-6, 73-78.

## About the authors

Miguel Ángel Alonso Rodriguez is an architect and surveying engineer. He holds a Ph.D. in architecture; his dissertation focused on the history of axonometry. He teaches descriptive geometry at the School of Architecture of the Polytechnical University of Madrid. He has also lectured in graduate courses on the history of axonometry and surveying in Madrid and at the Universidad Nacional Autónoma de Guadalajara. He has carried out surveys of historical buildings in archeological sites in Spain and Pompei with the support of both research grants, such as the ones included in *Cantería renacentista en la catedral de Murcia,* and commissions from such key Spanish institutions as Patrimonio Nacional and the Real Academia de San Fernando.

Elena Pliego de Andrés is a Spanish architect living and working in Madrid, Spain. She has had a professional practice since 1995. She teaches descriptive geometry at the School of Architecture of the Polytechnical University of Madrid, and Building Design and Construction in the vocational school IES Islas Filipinas in Madrid. Her research interests include Joseph Durm and his *Handbuch der Architektur*. She is currently developing a dissertation on Georg Gottlob Ungewitter's *Lehrbuch der gotischen Konstruktionen* and the Gothic construction techniques, under the advisement of Santiago Huerta.

Alberto Sanjurjo Álvarez is a Spanish architect living and working in Madrid. He has had a professional practice since 1991. He teaches descriptive geometry in the Polytechnic School of Architecture in the University San Pablo CEU, Madrid. He is currently developing research studies on the history of descriptive geometry and history of construction, especially stereotomy and stonecutting craft and techniques, while developing a dissertation on spiral staircaises under the advisement of Enrique Rabasa Díaz ("Las escaleras de caracol en los tratados de cantería españoles de la Edad Moderna y su presencia en el patrimonio construido hispánico: estudio geométrico y constructivo").

# Marta Salvatore

Department of History, Design
and Restoration of Architecture
University of Rome
"La Sapienza"
Piazza Borghese, 9
Rome, ITALY
marta.salvatore@uniroma1.it

Keywords: Stereotomy,
descriptive geometry, Amédée
François Frézier, Girard
Desargues, Alexis Claude
Clairaut, coupe de pierres,
stonecutting, stonecutters,
masonry architecture, quadric
surfaces, quartics, curves of
double curvature, right cone,
oblique cone, conic sections,
circular sections

## Research

# *Prodromes of Descriptive Geometry in the* Traité de stéréotomie *by Amédée François Frézier*

**Abstract.** Among the sciences involved in the theorization of descriptive geometry, the stereotomy played a prominent role. The knowledge of the theory of the surfaces of which bodies are made up, their plane sections, their intersections, the representation methods to control the design operations in the plane, is the core of stereotomic design. This makes the stereotomy the forerunner, in terms of theory and tools, of modern descriptive geometry. A seminal essay is the 1737-1739 *Traité de stéréotomie* by Amédée François Frézier. This work, published a few years before Monge's Géométrie descriptive, summarize the descriptive geometry's state-of-the-art in that period. Notably in the first book, Frézier publishes an original study about the intersections between quadric surfaces and the projective-geometrical properties of the fourth-order curves derived from them.

*On sait que les procédés plus ou moins ingénieux employés depuis longtemps dans la coupe des pierres et la charpente ont conduit Monge à la création de la géométrie descriptive. Les secrets d'ateliers réunis, classés, et surtout rattachés à un petit nombre de principes, ont constitué les éléments d'une branche importante des sciences appliquées. Aujourd'hui la géométrie descriptive est enseignée d'une manière indépendante des questions qui lui ont donné naissance; néanmoins, parmi ses applications les plus utiles, il faut citer le tracé des épures relatives à la coupe des pierres et à la charpente. La plupart des exercices à traiter se rapportent à l'intersection des surfaces*

[Frere Gabriel Marie 1877/1996: 376)].

## Introduction

As we all know descriptive geometry is a science that uses models to portray figures in space. However in recent years its role appears to have changed and now has more to do with discovering the rules that govern the construction of two-dimensional graphic models. In so doing, it runs the risk of losing sight of its original vocation which, up until the last century, had been to be an indispensable tool for understanding the geometric properties of solid figures in space.

These spectacular cognitive opportunities, which inspired the work of surveyors up until the early twentieth century, can now be developed further by computer science which allows us to create extremely accurate drawings directly in space. Three-dimensional modelling makes it possible to overcome the limits of graphic representation methods and reconsider descriptive geometry as a continually evolving discipline, updating those sciences that have played a major role in its history with a view to including topics that so far have been impossible to explore.

Nexus Network Journal 13 (2011) 671–699        NEXUS NETWORK JOURNAL – VOL. 13, No. 3, 2011    **671**
DOI 10.1007/s00004-011-0086-0; *published online* 22 November 2011
© 2011 Kim Williams Books, Turin

Stereotomy is one of these sciences. It demonstrates how the art of stonecutting is dependant on an inalienable link between the model, the drawing and construction. A topical contemporary concept considers this link to be the load-bearing structure of modern descriptive geometry. The search for the prodromes of descriptive geometry in stereotomy leads us to explore all contributions – from the work done in stonecutters' workshops, to the Renaissance treatises and the geometric and mathematical works of the Age of Reason – that turned the art of stonecutting into a genuine science, thanks to the elaboration of a geometric theory that governed its practical application.

One of these contributions is the monumental work written by Amédée François Frézier [1737-1739] in the fist half of the eighteenth century, the last treatise written before the rationalisation developed by Gaspard Monge.[1] The *Traité de stéréotomie*, which appears to be a treatise on solid geometry, systematises and rationalises a cognitive process over two thousand years old.

## Stereotomy: art and science

A short review of certain decisive moments in the history of stereotomy can help clarify the geometric, mathematical and constructive knowledge available to Frézier at the time and explain the importance of the methodological contributions and contents of his treatise (for more information, see [Salvatore 2008]).

In 1737 Frézier wrote *La théorie et la pratique de la coupe de pierres et de bois pour la construction des voûtes, et autre parties des bâtiments civils et militaire, ou traité de stéréotomie à l'usage de l'architecture* (The theory and practice of stonecutting and wood working to build vaults, and other parts of civilian and military buildings, or the treatise of stereotomy in architecture). The treatise appeared at the end of a gradual process during which stonecutting developed into a true science. This evolution was driven by a cross-fertilisation between different fields of learning, much like what happened with the different disciplines used by stereotomy, which include geometry, mathematics, statics, construction, etc.

Like all evolutionary processes, it is impossible to say when it started, but we do know that certain epochal changes contributed to shifting the focus from the actual practice of stonecutting to the development of a universal theory.

Improvements in architectural design and therefore in the role of architects during construction is just one such change. It was a gradual process, during which building became subordinate to planning and design. The architectural project as we know it today can be considered one of the most "fertile" inventions of the Renaissance, which inspired a cultural, speculative approach to architecture until then reserved only to science [Potié 1996: 8-12]. Design became more important than the actual building process; it attracted and coalesced technical disciplines and all those fields in which the conception and construction of a work is based on an intelligent use of the graphic method behind these operations: double associated projection.[2] Renaissance architecture testifies to the advanced use of double associated projections, a trend that started at that time. However, as Dürer rightly observes, it is possible that the graphic method was developed and improved upon by stonecutters[3] and that anyone wishing to study proportions must first have assimilated the way in which measurements are taken and understood how everything has to be arranged in plan and elevation, according to the method always adopted by stonecutters [Peiffer 1995: 59].

Improvements in both graphic methods and stonecutting techniques seem to have reciprocally influenced and fertilised both fields over a period of time. In fact, developments in the graphic method led to gradual improvements in stonecutting; likewise, building requirements slowly gave rise to redoubled efforts to invent new instruments to control geometric forms.

However, other factors were crucial in this transformation of stonecutting into a science: the relationship between the evolution of graphic methods and the contributions which increased the understanding of the geometry of forms. The period when De l'Orme published *Le premier tôme de l'architecture*, in which he developed the *traits géométriques* needed to build several kinds of lithic architecture corresponds to what became known as the Renaissance of geometry. The dissemination of the works of Alexandrian mathematics and Renaissance research in mathematics laid the groundwork for the fertile scientific work of the seventeenth and eighteenth centuries. The geometric and mathematical studies of certain surfaces and plane sections published at that time provided more general knowledge about the complex forms of space. However, some people did try to generalise even about the use of *trait*,[4] which despite its declared intention proposed different solutions for each of the examples described. One such person was Girard Desargues, who in his *Brouillon projet* developed a universal building method based on theory. He demonstrated that abstract geometric logic, rather than practice and experience, was the best way to identify where to cut the stone, thereby robbing master stonecutters of part of their work. The search for the theory behind an applied science was a constant in Desargues's work in the mid-seventeenth century, as it was for Frézier, who in the fourth book of his *Traité de stéréotomie*, in the chapters dedicated to the problems of building a barrel vault, cites a study on stonecutting in the *Brouillon projet* entitled "Concise explanation of the Desargues method" [Frézier 1737: 191-206]. Frézier believed that Desargues had sensed that all the *traits* about the construction of barrel vaults, straight, skewed, sloped or slanted could be reduced to just one problem – identifying the angle between the axis and a generic section of a cylinder [Poudra 1861: 305-351]. We know that Desargues's theories were not popular among his contemporaries,[5] but in the world of stereotomy the search for a universal theory for all Desargues's work gave rise to two parallel schools of thought. One school believed in practice and included Jacques Curabelle, Mathurin Jousse, François Derand and Jean Baptiste de la Rue, who were indifferent to the search for general principles and whose works are in fact extensive collections of ad hoc solutions for each case in question. The other school comprised the theorists, such as Philippe de La Hire, a pupil of Desargues and Amédée François Frézier [Laurent and Sakarovitch 1989].

The heterogeneous origins of contributions to the science of projection developed between the sixteenth and eighteenth centuries is summarised by Jean Nicolas Pierre Hachette in his introductory speech to the class of Physics and Mathematics in 1812 [1816]. Hachette describes how the scope of descriptive geometry is to represent three-dimensional objects on plane surfaces and how the way to achieve this consists in projecting these objects onto these planes. The science of projection in general can therefore be divided into two fields: one involves the reasoned but strictly graphic execution of these projections while the other involves purely analytical theory. Hachette goes on to say that these two sciences are simply two different ways of tackling the same problem, and that even if these procedures appear to have very little in common, the constant congruence in their results generates a continuous rapprochement that should not surprise us. This correspondence shows that what appears at first glance to be two very different theories in fact provide reciprocal confirmation of each other's validity, explanations and generalisations:

> *l'une, en un mot, former des tableaux qui parlent aux yeux, tandis que l'autre s'occupe à les décire aussi fidèlement qu'exactement dans la langue qui lui est propre* [Hachette 1816: 235].

Hachette goes on to say that the science of projection had been enriched by graphic contributions, for instance by De l'Orme, Jousse, Deran or De La Rue, who had shed light on the art of *trait,* and by analytical theories which had no direct application in art, for example those of Alexis Claude Clairaut, the first to write a treatise on curves of double curvature. In his *Traité de stéréotomie,*[6] Frézier was the first to consider both aspects of the science of projections – graphics and analysis – in conjunction. Frézier's approach was quite uncommon, because mixed methods presumed a rather advanced understanding of the theory of calculus, but this is exactly what Frézier did: he was intelligent enough to collate the studies on the theory of calculus and use them in support of the science of stonecutting. One of the most important mathematical contributions known to and cited by Frézier were the studies on the curves of double curvature,[7] in particular the *Recherches sur les courbes à double courbure* by Alexis Claude Clairaut [1731].[8] The ancients were not unaware of the curves of double curvature, and contemporaries also studied these curves.[9] However as Clairaut himself states in the preface to the *Recherches sur les courbes à double courbure,* with the exception of several studies by Desargues, it was the first written work in which all aspects of these space curves were described. In Clairaut's work, every curve is described inside a solid right angle (or is simply inserted in a three-dimensional Cartesian system) and is defined by two of the three plane curves created by the orthogonal projection of the curve of double curvature onto planes which define the solid angle. Clairaut projected all the points of the curve on three respectively orthogonal planes, as was customary for plane curves in a two-dimensional system. This gave him three curve projections, from whose equations, chosen pairwise, he obtained the unknown of the curve of double curvature.[10] Even if Clairaut's objective as a mathematician was to determine the equation of a curve of double curvature, the proof of the algebraic conclusions was based on a clear geometric analysis of the problems, starting with the premise that any curve of double curvature can be considered as being derived from the intersection of two or more curved surfaces. All geometric considerations in support of the theory of calculus are illustrated in the *planches* annexed to the treatise (figs. 1, 2).[11]

The improvements in the method of double associated projections, the search for a universal geometric theory by the school of Desargues, and progress in mathematical studies on curves, surfaces and their intersections were the humus in which Monge's descriptive geometry was to grow, the same humus that persuaded Frézier to systematise the discipline at a time when, as the author himself writes in the second introduction to the treatise: *Je sçai qu'aujourd'hui la Geometrie Lineaire n'est plus gueres à la mode, & que pour se donner un air de Science, il faut faire parade de l'Analyse* ... [Frézier 1737: IX].

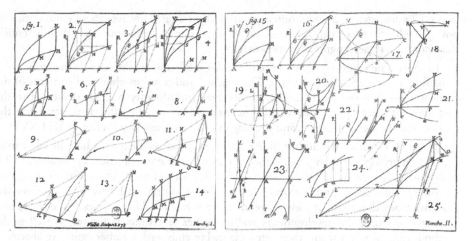

Fig. 1. *Planches* I and II from *Recherches sur les courbes à double courbure* by Alexis Claude Clairaut [1731]

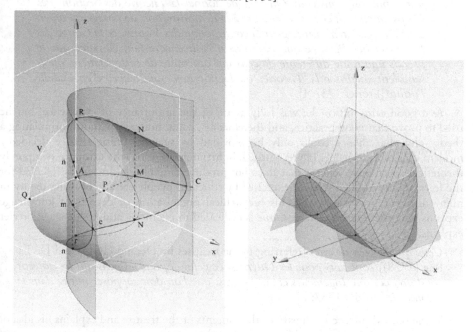

Fig. 2. Conics projection of a quartic; reconstruction of figure 17 in the second *planche* from *Recherches sur les courbes à double courbure* by Alexis Claude Clairaut [1731]

## *The* Traité de stéréotomie *by Amédée François Frézier*

Amédée François Frézier was one of the most respected figures in the history of stereotomy, yet studies of his work are few in number and very patchy.[12] One of the most cultured figures of the eighteenth century in France, Frézier's fame did not depend only on stereotomy.[13] However, critics acknowledge that Frézier's work was the first structured contribution to the art of stonecutting. Evidence of this comes in the form of papers by the most illustrious theorists of descriptive geometry, including Hachette,

Loria, Chasles, etc., and later contemporaries like Taton. However these same critics do not provide detailed explanations as to the merits and limits of the "scientific stereotomy"[14] they recognised in his works.

In short, one could say that compared to his predecessors, Frézier was the author of a bona fide treatise on solid geometry. The unusual and novel method and contents of the work was summarised by the author in his three introductory prefaces dedicated to the usefulness of theory in the arts which interest architecture, the subjects illustrated in the treatise and their classification, and the origins of stereotomy [see Frézier 1737: I-XVI].

In his first preface the author illustrates his methods and explains how he constantly focused on trying to find a general theory that is a guide for practical applications. This distinction between theory and practice is the backbone of the treatise. The work focuses on purely theoretical propositions which tackle building problems that arise on the worksite from the point of view of geometric abstraction. He analyses the surfaces and their intersections, the developments and properties of geometric figures, as well as the methods needed to represent them on a plane. He ends with a rather extensive abacus of experiments and practical applications:

> *Je me propose dans cette Ouvrage de donner la Theorie des Sections des Corps, autant qu'elle est necessaire à la démonstration de l'usage qu'on en peut faire en Architecture peur la construction des Voutes, & la Coupe des Pierres et des Bois, ce que personne n'avoit encore fait; & parce que je prends une route differente de ceux qui ont traité de cette Matière, qu'ils semblent mépriser la Théorie, ou l'ignorer: je vais tacher d'en établir l'utilité* [Frézier 1737: I].

As a good army officer he was fully aware of the importance of experience, but he tried to prove that good practice, and therefore art, could not exist without formulating a theory and that theory could only be provided by science. Frézier had a decidedly modern idea of knowledge. He recognised, in hierarchical order, the existence of a purely theoretical speculative science that does not produce tangible results and is reserved for the learned, and applied science, exercised by educated people capable of using theory in many different practical ways. This hierarchical organisation of degrees of knowledge explains why the *Traité de stéréotomie* is dedicated to engineers and architects.[15] Frézier explains:

> *Ceux que je vien de citer* [referring to the treatises by Derand (1643) and De La Rue (1728)] *sont faits pour les Ouvriers, & celui-ci pour les gens qui les doivent conduire, comme Ingenieurs et Architectes, que l'on doit supposer initiez dans la Géometrie* [Frézier 1737: I-XVI].

In the second preface he illustrates the contents of the treatise and explains his idea of stereotomy, which he says should not be called the art of the artisan who cuts stones, but instead a mathematical science that should guide their hand when assembling the various parts to create a single body, parts which are arranged in such a way that it is their weight that provides reciprocal support, without the use of mortar or concrete.[16] Stereotomy depends on geometry, which is crucial if one wants to understand the shape of objects and also mechanics and statics. Frézier noted that many treatises had been written by contemporaries on the principles of statics and mechanics applied to buildings and refers any reader interested in this subject to the works of some of the most famous mathematicians of the day.[17] However, the way to use geometry in construction had not yet been published in a treatise and although there was extensive scientific knowledge about geometry, it was rarely used in practice. This is why the treatise focused only on the study of the geometry of forms, in particular the forms that occurred repeatedly in buildings.

Having said this, Frézier outlines the structure of the book, which was divided into volumes and books following a methodological order in which he included all aspects of this discipline, rigorously dividing the subject matter according to different fields of learning. The first volume includes the first three books illustrating the theoretical element of the discipline, in other words, the *science of stereotomy*; the second and third volumes, which together form the fourth book, are dedicated to the way in which the propositions in the first book can be used, in other words to the *art of stereotomy*. In particular:

- Book I is dedicated to *Tomomorphie*, or the "form of sections," needed to understand the nature of the plane or space curves created by the intersection of solids;
- Book II is dedicated to *Tomographie*, or the "drawing of sections," needed to draw the above-mentioned lines on flat surfaces;
- Book III is dedicated to *Stereographie*, the representation of solids and their sections on plane surfaces using: double associated projections, *Ichnographie* and *Ortographie*; their development on a plane, *Epipedographie*; and our knowledge of the angles created by the surfaces of solids, *Goniographie*;
- Book IV is dedicated to *Tomotechnie*, the art of construction (especially vaulted systems), by applying graphic models to stone.

In his treatise Frézier seized every opportunity to point out the mistakes made by his predecessors. By comparing his own work with previous publications, in the third introductory preface he illustrated the novel elements of his treatise for the reader, in particular his research on theory, his corrections of the mistakes made by others, his proposed novel solutions, including several examples.[18]

One should bear in mind that Frézier's background in geometry was very similar to Desargues's. Progress in the field of mathematics over the course of the entire seventeenth century and the many ways in which algebra was applied ended by relegating geometry to the back burner. Algebra became the instrument of choice for solving problems, and the works by Desargues, de La Hire and Pascal were forgotten: the eclipse of geometry was to last more than a hundred years [Kline 1991: 456-465]. The advent of infinitesimal calculus heralded a new concept in mathematics that captured the attention of most mathematicians. Many initially remained faithful to old geometry, and in fact it was Liebniz who noted that geometricians could often demonstrate in a few words what would take much more time using calculus [Coutrault 1903: 181] but that the powerful resources provided by infinitesimal calculus induced mathematicians to make new speculations [Chasles 1837: 142]. Despite all this, Frézier was well aware of the subordinate role to which geometry was condemned during his lifetime. This is evident in his introduction to the treatise:

> ... *l'ancienne Geometrie ... fournit à la nouvelle des fondaments solide, paticulierrement dans la matiere dont il s'agit, où le calcul Algebrique ne pourroit être utile qu'entre les mains de ceux qui y sont plus avancez, que ne le sont ordinairement la plûpart des gens qui se mêlent d'Architecture .... D'ailleurs elle conduit plus naturellement à la pratique du Traits de la Coupe des Solides, & fait selon moi plus d'impression dans la mémoire, où les Surfaces & les Lignes se gravent plus profondement que les préceptes des formules Algebriques* [Frézier 1737: X].

Frézier's work is one of the few eighteenth-century contributions to geometry we can define as being developed "the old way" (we shouldn't forget he learnt what he knew about mathematics from Philippe de La Hire). Compared to his predecessors, Frézier had all the technical knowledge and hands-on experience to turn a long searched-for chimaera into reality.

## *The first book:* tomomorphie

In the first book on *tomomorphie*, in other words the study of the form of the sections of bodies, Frézier illustrates the theory of plane lines and curves of double curvature created by intersecting surfaces.

The treatise focuses in an abstract manner on quadric surfaces, especially those which recur more often than others in the architecture of vaults illustrated in Book IV. Frézier starts with a concrete example so that the reader can understand why comprehension of the section of bodies is necessary in stereotomy. He placed half a melon on a flat surface and sectioned it in different ways, slicing or cutting it by rotating a knife around the tip of the blade. Plane surfaces create the simplest sections, curved surfaces the most complex, proving that geometry is necessary to understand them [Frézier 1737: 2-4]. Based on the comparison between the different sections of the melon, Frézier defines two types of sections: plane sections, created by the intersection of quadric surfaces with a generically-positioned plane; and space sections, created instead by the intersection of two quadric surfaces.

The first book is divided into two parts. The first focuses on the plane sections of quadric surfaces which are conic sections; the second is dedicated to solid intersections,[19] in particular to the curves of the fourth order and to special cases in which the intersection of two quadrics is a conic.[20] In each of these sections, each intersection is analysed with the same thoroughness and method that characterise in the whole treatise. Frézier starts with a general introduction to plane sections, followed by a systematic analysis of the plane sections of spheres, cones, cylinders and the surfaces he defines as 'regularly irregular',[21] such as spheroids,[22] conoids, ellipsoids, rings and helicoids (fig. 3).

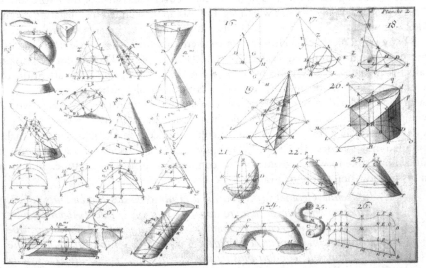

Fig. 3. Conics created by intersection of a quadric surface with a plane; *planches* I and II from *Traité de stéréotomie* by Amédée François Frézier [1737-1739]

As Frézier himself admits, his approach to conic sections is not at all novel compared to previous works on this subject, but he includes it so that the reader (who he imagines is acquainted with the rudiments of geometry) does not have to consult other books. Frézier refers here to Apollonius, but also mentions the works by Philippe de La Hire and the Marquis de l'Hôpital.[23] One should not forget that beginning in the seventeenth century, the work by Desargues and his pupils had contributed innovative treatises on conic sections[24] to the field of science. In fact, when Frézier wrote his treatise, people were already familiar with the projective properties of the circle and the possibility to obtain the many properties of conics from this figure. Furthermore, studies were so advanced that the method used to turn a circle into a cone on a plane, used by de La Hire and Le Poivre, preempted the properties of homologous figures relating to the correspondence between points and straight lines of a circle and a conic, theorised later by Poncelet in his *Traité de propriétés projectives* (1822).[25]

Like the two treatises which so obviously inspired him, Frézier studies the properties of conics as curves obtained from the plane sections of quadrics and as graphic lines of the plane. As mentioned earlier, although he did not propose anything new, his studies on the circular sections of cones and quadric cylinders and the analysis of the plane sections of hollow bodies are worthy of note and consideration.

The problem of finding the circular sections of a quadric cone was solved by Apollonius who used algebra on an oblique circular cone. It was Father Mersenne who pointed out that this problem was reproposed by Girard Desargues, capturing the attention of several scholars, including geometricians and mathematicians [Mersenne 1644: 331]. Desargues invited his colleagues to generalise the contemporary concept of conics defined as sections of a cone with a circular base. He asked himself whether a generic quadric cone, regardless of what conic section was used as a base, could be cut by a plane and produce a circular circumference and, if so, what method should be used to determine the position of the plane section [Taton 1951: 42-43]. At that time, this was a complicated question, to which Descartes proposed an analytical solution based only on a cone with a parabolic base [Chasles 1837: 82-83]. Mersenne goes on to say that it was Desargues who formulated a geometric solution to the problem based on the construction of the main axis of the cone, but unfortunately no proof remains [Taton 1951: 42].

Frézier also proposed a solution to the problem, looking for circular sections in a quadric cone and cylinder; it was a geometric contribution which, in the case of cones, is misleading, but it should not be forgotten because finding the solution to this problem is closely linked to finding the main axes of quadric surfaces.

Like de La Hire and the Marquis de l'Hôpital, Frézier considered a conic surface to be a surface created by moving a straight line, fixed at a point (the top of the cone), around any circumference in space. He uses the word 'axis' to indicate the segment joining the centre of the base circumference to the vertex of the surface. He admits that right cones and oblique cones exist, depending on whether the axis is perpendicular or oblique to the plane of the circular base. Finally, he senses that an oblique circular cone is identical to a right elliptical cone and for this reason calls it right cone on its base (fig. 4).[26] If he had focused more on the fact that only two kinds of conic surfaces exist, i.e., two types of right cones whose only difference is the solid angle at the vertex, this might have had a positive impact his search for the three main axes of the quadrics. In actual fact, Frézier was still far from formulating this concept and this is one of the reasons why there are several inaccuracies in the way he determined circular sections.

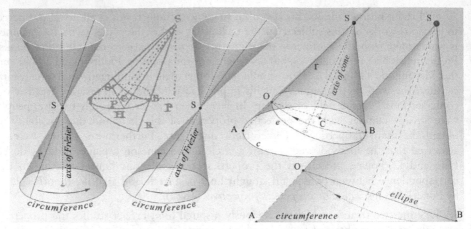

Fig. 4. Reconstruction of the genesis of right cones and oblique cones described by Frézier and the possibility of obtaining from an oblique cone the corresponding right cone on its base; the drawing in grey is number 10 in the *planche* I from *Éléments de stéréotomie à l'usage de l'architecture pour la coupe des pierres*, written by Frézier in 1760

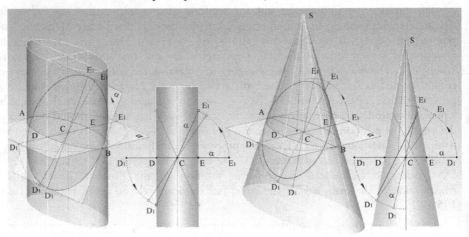

Fig. 5. Reconstruction of Frézier's method to determine circular sections in a quadric cylinder and in a quadric cone

Given an elliptical cone and cylinder sectioned by a plane that together form any elliptical section, Frézier makes the ellipse rotate around one of its axes (fig. 5). During rotation, the other axis changes its length until it becomes as long as the first. The congruence of the two axes determines the position of the plane that sections the quadric through a circular circumference (of which the two axis become the radii). This is an effective method for elliptical cylinders because the centre of the two section circumferences, like the centre of any elliptical section, lies on the main axis inside the quadric. This is not the case for cones: here the centres of the circular sections, due to the obliqueness of the generatrices compared to the axis, do not lie on the main axis but are symmetrical to it and therefore cannot coincide with the centre of a generic ellipse section or with that of the right section of a cone (fig. 5). Frézier was not familiar with the concept of main axes of quadrics. In fact, while the main axis of cylinders coincides with the segment he calls an axis, the same is not true for cones. Even though he sensed it was possible to turn oblique cones into right elliptical cones and had therefore identified the main axis inside the quadric, this perception is not developed in the treatise.[27]

The circular sections in a quadric cone were determined in the first half of the nineteenth century by Théodore Olivier, who used a right elliptical cone in his generalisation [see Olivier 1852: 199-202]. Olivier noted that the circular sections of a right elliptical cone are produced by the intersection of the cone with a sphere whose centre coincides with the centre of the base ellipse and whose radius is equal to the distance of the centre of the ellipse from one of the generatrices of the cone passing the end of the major axis of the base ellipse (fig. 6).

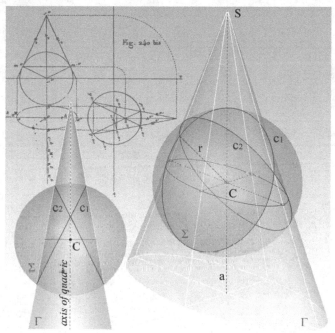

Fig. 6. Théodore Olivier's method for determining circular sections in a right cone on an elliptic base; the original drawing comes from the *planche* 75e of *Cours de géométrie descriptive* (1852)

Since in every oblique cone the main axis inside the surface passes through the centre of a elliptical section perpendicular to it – in other words, every oblique cone is a right elliptical cone – it is possible to apply Olivier's solution to any quadric cone. To obtain a right cone from an oblique cone, it is necessary to know the main axis inside the quadric, but the method to create the main axes of these surfaces is provided by mathematical analysis using procedures that could not be represented graphically with the tools available to geometricians before the second half of the twentieth century.

Today the problem can be solved differently. As specified in the introduction, it is one of those cases in which by applying representation methods to computerised methods, solutions can be found directly in space. Determining the axes of a quadric cone (or quadrics in general) can be geometrically solved very quickly by mathematical modelling,[28] using the properties of the barycentre of solid figures, based on this observation: the main axis inside a quadric is the geometric locus of the barycentres of solid figures made up of the quadric and by planes perpendicular to the axis.

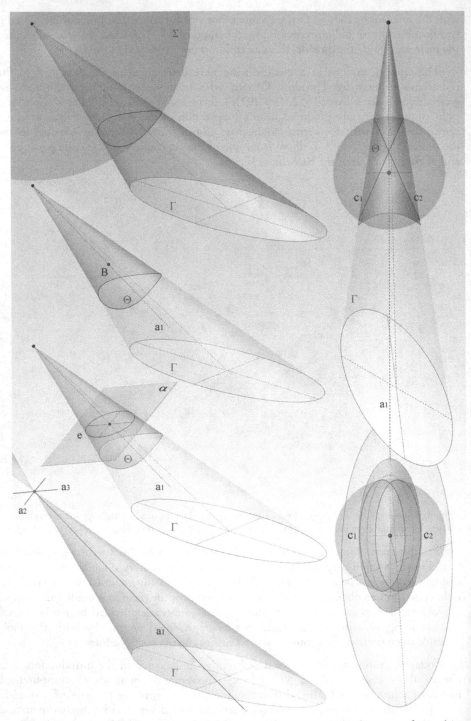

Fig. 7. Properties of the barycentre of solid figures used to construct the three axes of a quadric cone; generalization of Olivier's method to construct the circular sections in a right cone on an elliptic base extended to any quadric cone

In the case of a generic oblique cone, given the barycentre of any corresponding right cone, we will know the direction of the main axis we are looking for. Since it is impossible to obtain the corresponding right cone from an oblique cone without knowing the direction of the main internal axis – which is in fact what we are trying to establish – we will have to use a figure with characteristics of symmetry similar to those of the right cone; this can be achieved by building a sphere with any radius whose centre lies at the top of the cone. The curve intersection of the sphere and the cone is a curve of the fourth order. Eliminating the part of the cone outside the sphere will give a solid figure made up of the conic surface and part of the spherical surface, the barycentre of which (the computer calculates this automatically) lies on the unknown axis of the quadric, thanks to the above-mentioned properties of symmetry. The straight line between the barycentre of this figure and the vertex of the cone is the main axis. To find any right cone starting with an oblique cone, the quadric has to be sectioned by a plane perpendicular to the main axis (fig. 7) [Salvatore 2008: 139-146].

Frézier's considerations about hollow bodies, in the first part of the first book of the *Traité de Stéréotomie*, are also worthy of note. In stereotomy, the theory of conic sections should also apply to hollow bodies, because the forms used in masonry architecture are similar to hollow bodies, particularly to hollow cones and cylinders, the forms most used in vaulted systems. Frézier demonstrates how the plane section of a hollow circular cone of uniform thickness is an elliptical ring made up of two ellipses that cannot be concentric. He also demonstrates how the plane section of a hollow circular cylinder is an elliptical ring made up of concentric but not equidistant ellipses, because two concentric ellipses are never equidistant (fig. 8).

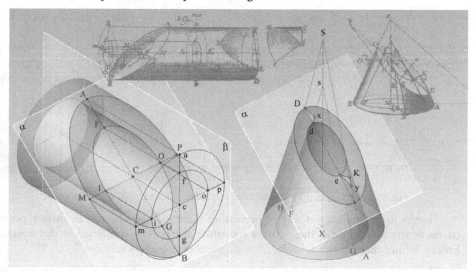

Fig. 8. Plane sections of a hollow cylinder and a hollow cone, whose right sections are circular rings

He then goes on to apply this to oblique bodies, treading on thin ice in the cases where it is not always possible to prove certain theories. Frézier believes that there's only one case in which a hollow sectioned cone produces an elliptical crown made up of concentric ellipses: when the cone is scalene and the elliptical ring is given by the section orthogonal to the axis which, he goes on to say, is like considering it a right elliptical cone. However, the method used by Frézier to determine a right elliptical cone starting with an oblique circular cone cannot be used for hollow cones, because the surfaces of a hollow oblique cone whose base is a circle cannot be equidistant. So, if we wanted to obtain from this hollow cone the corresponding right cone, we would see that the right sections of the two surfaces have different positions because the main axes of the two quadric surfaces are different. With reference to hollow oblique cylinders with equidistant surfaces, the sections will generally be concentric ellipses, except for one case in which the section will be a circular ring.[29]

The second part of the treatise, containing the most innovative contributions for that period, focuses on the *curves of double curvature* created by the intersection between quadric surfaces. Frézier was inspired by Clairaut's *Recherches sur les courbes à double courbure* [1731] published a few years earlier. Clairaut mathematically analysed the curves of double curvature, while Frézier adopted an approach which was rather unusual for that period: geometry. Frézier was interested in understanding curves of double curvature in order to control all aspects of a vaulted system morphologically; this meant understanding the geometric nature of the curves created by the intersection of quadric surfaces. The treatise also explains how to apply this knowledge in practice. Each time Frézier analysed an intersection, he backed it up with concrete examples showing how important it was to understand the problems of intersection. In the first part of the first book, plane sections were used to explain simple vaults; likewise curves of double curvature, intersection of quadric surfaces generally make up the corners of composite vaults, except for several special cases in which these intersections give rise to plane curves (fig. 9).

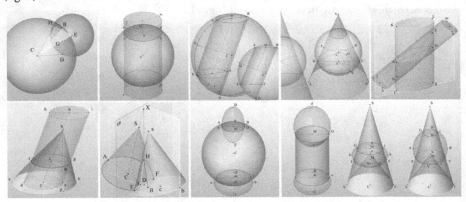

Fig. 9. Conics created by intersection of quadric surfaces variously juxtaposed

Thanks to his meticulous way of presenting these examples he reveals three types of curves of the fourth order; these curves constitute the novelty contained in this treatise. Frézier defines these curves as *cicloïmbre, ellipsimbre* and *ellipsoïdimbre*.[30]

The treatise focuses on the possible intersections between bodies in positions that are always different; the subject is tackled with the same meticulousness Frézier uses to explain plane sections (figs. 10, 11, 12). He analyses the intersection between spheres; between cylinders and cones; between cylinders and other cylinders and cones; between cones; between spheroids and spheres; and between cylinders and cones (according to the order shown in the synoptic table, fig. 13).

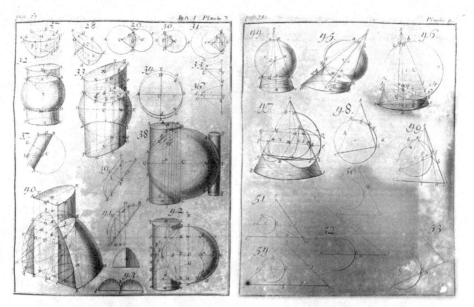

Fig. 10. Conics and quartics created by intersection of a sphere with another quadric surface; *planches* III, IV from *Traité de stéréotomie* by Amédée François Frézier [1737]

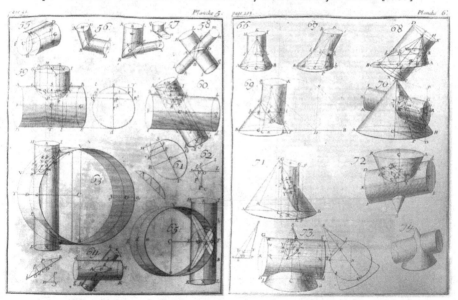

Fig. 11. Conics and quartics created by intersection of a cylinder with another quadric surface; *planches* V, VI from *Traité de stéréotomie* by Amédée François Frézier [1737]

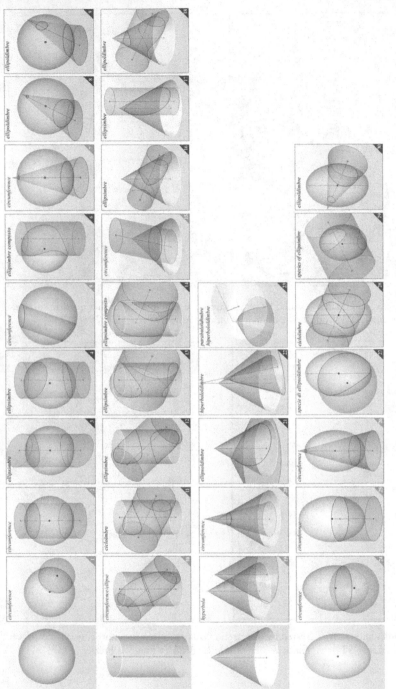

Fig. 13. Synoptic table of the examples of intersection between quadrics analyzed by Frézier, following the order of the original treatise; the colours below right show the type of curve of the fourth order related to each example

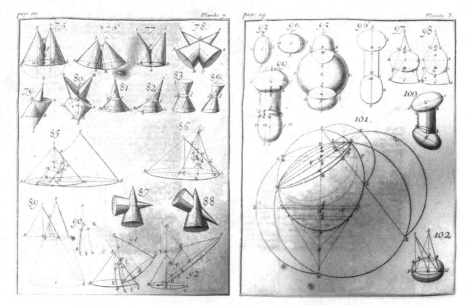

Fig. 12. Conics and quartics created by intersection of a cone and a spheroid with another quadric surface; planches VII, VIII from *Traité de stéréotomie* by Amédée François Frézier [1737]

Frézier wanted to provide his readers with a concrete example of these curves, so he suggested they imagine a circle on the spine of a book about to be bound and inserted in the press by the bookbinder. The reader was asked to imagine what would happen when the bookbinder rolled the spine: the pages would slide over one another, causing the pages to move, less in the centre and more at the edges. Thus the circle would no longer be a circle, but instead the curve of fourth order that Frézier called a *cicloïmbre*, which belongs, not to a plane, but to a cylindrical surface (fig. 14). The *cicloïmbre* is therefore a quartic, obtained from the parallel projection of a circle on a cylindrical surface. Likewise, the parallel projection of an ellipse on any quartic surface is called an *ellipsimbre*,[31] while the central projection of a circle or an ellipse on any quadric surface is called an *ellipsoïdimbre* (fig. 15).[32] These are obviously curves with two branches: Frézier shows both branches, but for reasons of symmetry, analyses only one.

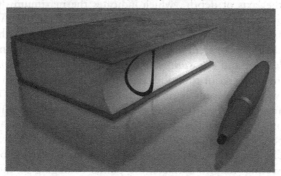

Fig. 14. Example selected by Frézier to explain the geometric nature of the curves of double curvature; if you draw a circle on the spine of a book about to be bound, the rounding of the spine will transform the circle in a curve of double curvature, specifically a *cicloïmbre*

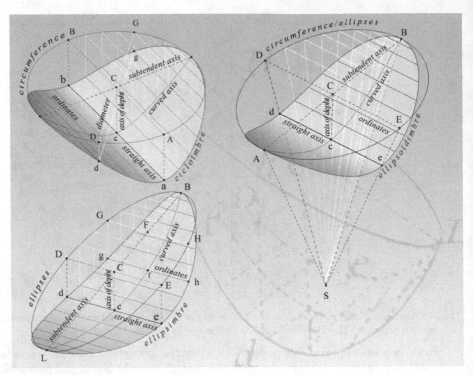

Fig. 15. The *cicloïmbre*, the *ellipsimbre* and the *ellipsoïdimbre*, the three types of curves of the fourth order identified by Frézier in proportion to the conic from which they originate through parallel or central projection

These curves, when projected using a direction of central or parallel projection on a plane surface, are projected into the conics. If we consider that these quartics are obtained through parallel or central projection of the conics on quadric surfaces, during projection some of the more characteristic lines of the conics which produced them will remain unaltered, while others will change shape.[33]

Since these are curves of double curvature, a segment will establish its depth, taking its name from the 'axis of depth', which indicates the maximum distance of the solid section[34] from the plane of the conic that generated the quartic. As per plane curves, Frézier defines the segments between two opposite points of a curve that pass through the axis of depth as diameters of a curve of double curvature; they are identical to those of the conic in the *cicloïmbre* and *ellipsimbre*, and parallel in the *ellipsoïdimbre*. He defines the straight axis as the diameter of the solid section which coincides with the genetratrices of the surfaces on which the conic is projected, and subtended axis as the diameter of the conic passing through the points where the curve of double curvature is tangent to the plane of the conic. The curve that halves the solid section and lies on the plane passing through the subtended axis perpendicular to the plane of the conic, is called a 'curved axis' (fig. 15). The projections linking the conics to the quartics allow Frézier to locate, given two intersecting quadric surfaces, the points belonging to the intersection space curve.

It is possible to pass any number of surfaces through every curve of double curvature, especially two quadric surfaces, or more precisely, those from which one imagines the curve was created through intersection. In the projective method proposed here to describe curves of the fourth order, one of the two quadric surfaces is always a cone or a cylinder whose generatrices are the straight lines projecting the conic, while the other is the quadric surface to which, according to Frézier, the solid section belongs (i.e., the part of the surface defined by the curve of double curvature).

Note that the curved axis of a curve of double curvature is a conic, the plane section of one of the quadric surfaces that intersect and produce the quartic, but it is also the plane section of a third quadric surface that passes through the same space curve. Clairaut's studies had shown how a curve of the fourth order is projected onto three triorthogonal planes in a conic and how there are at least three quadric surfaces [see Clairaut, 1731: 29-30] that belong to the curve of the fourth order. This can be demonstrated in geometry considering that a curve of the fourth order can always be projected on three respectively orthogonal planes in three conics when regular quadric surfaces which intersect have, pairwise, common planes of orthogonal symmetry.[35] Frézier is familiar with Clairaut's work and therefore knows that a third quadric surface of the space curve exists, but does not examine it in his treatise. Instead he chooses more or less convenient solid sections depending on the intersection (fig. 16).

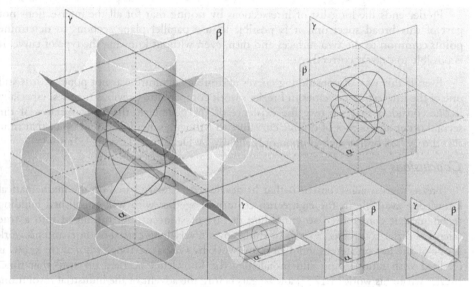

Fig. 16. Planes of orthogonal symmetry common to three intersecting quadrics

In short, the intersection of quadric surfaces can create three different kinds of curves of double curvature, the *cicloïmbre*, *ellipsimbre* and *ellipsoidïmbre*, which can always be projected, either in parallel or central projection, in the conics and which, like the conics which create them, have characteristic axes and lines. This kind of classification of the intersections between quadrics is the "load-bearing" structure of the second part of the first book. After having introduced the concept of curves of double curvature and illustrated the difference between them, Frézier systematically tackles the possible combinations of intersections between surfaces whose morphological variations and position determine different kinds of curves of the fourth order. There are many variables in intersections between quadric surfaces, which, when combined, influence the type of space curve created. Therefore parameters such as the reciprocal position of the bodies, especially the position of their axes and centres, or the straight or oblique nature of the surfaces and the type of intersection, are all important because all these parameters determine the number of generatrices of the quadric sectioned during intersection.

It is just as important to establish which quadric crosses the other, because the generatrices of the former become the projective straight lines. The *cicloïmbre* and *ellipsimbre* are curves that occur when in the intersection between quadrics and the cylinders cross another surface, because the generatrices of the cylinders, parallel by construction, are considered projective straight lines and because the plane sections of the round or elliptical quadric cylinders are always circles or ellipses (figs. 17, 18). Likewise, the *ellipsoïdimbre* occurs in all those intersections in which the cones cross another quadric. In this case, the two branches of the curve, the one entering and the one exiting, will be the same kind of curve, but each will be different because they are projected in a conic by central and not parallel projection (fig. 19). To these three main types we must add another two believed to be variants of the *ellipsoïdimbre*, and defined respectively as *paraboloïdimbre* and *hiperboloïdimbre*, which refer exclusively to intersections between cones, so the intersection space curve is projected onto a plane through central projection in a parabola or a hyperbola (fig. 20).

Frézier ends the long list of intersections by noting that for all the intersections not part of this broad spectrum, it is possible to use parallel plane sections to determine points common to the two surfaces and then, even without knowing the type of curve, it is possible to draw it correctly.

Even if this way of considering curves of double curvature was not popular, it is still one of the interesting novelties in Frézier's work compared to previous treatises, especially if one considers that we owe the existence of the geometric theory of surfaces of the second order and curves of double curvature to Monge's school, and that up to that time this theory had been touched upon only slightly by Desargues [Loria, 1931: 80-81].

## Conclusions

Frézier's great achievement is that he developed a theory based on the mathematical knowledge available in the eighteenth century. Together with well-established building techniques, it contributed to ennobling the art of stonecutting, turning it into a true science. Frézier completed a series of studies over a very long period of time, and his work is one of the last on stereotomy before the work by Gaspard Monge who, as we know, is credited with rationalising and broadening the knowledge accrued up until that time. Frézier wrote his work in 1737, a few years before the advent of the industrial revolution, at a time in history in which production and construction were to change forever and would no longer be the protagonists of a long but constant transformation. Monge based his search for a general theory on the need to satisfy the new requirements of industrial production. Similarly it was the search for a universal theory and its dissemination amongst professionals that fired Frézier's work, giving it an extremely modern flavour. Together with its novel contents, the work is an important landmark in the history of the science of representation.

To review and consider descriptive geometry as a discipline that is forever changing means updating and reviving those disciplines which, although considered dead and gone, have contributed to its history. Seen through modern eyes, stereotomy now appears to be a precious resource for the modern science of representation.

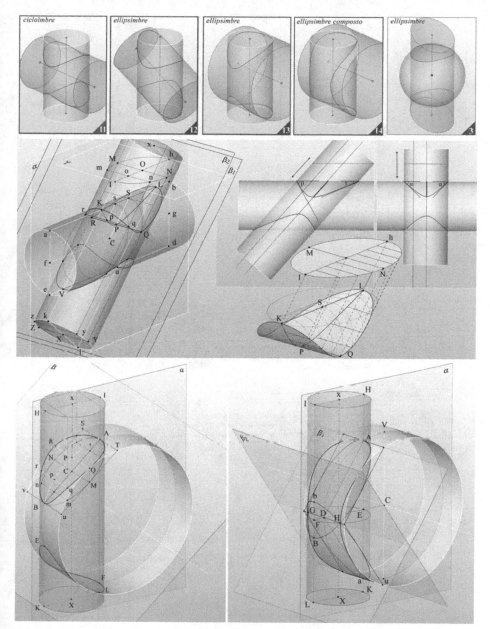

Fig. 17. Examples of intersection between quadrics, where the curve of the fourth order is a *cicloïmbre* or an *ellipsimbre*; specifically, reconstruction of the examples of intersection regarding two cylinders

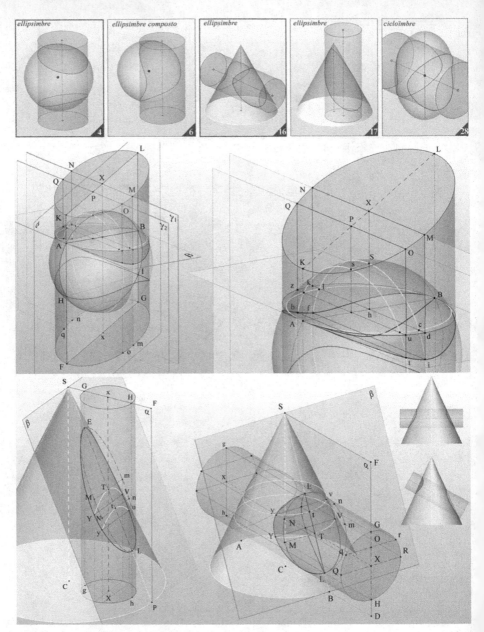

Fig. 18. Examples of intersection between quadrics, where the curve of the fourth order is a *cicloïmbre* or an *ellipsimbre*; specifically, reconstruction of the examples of intersection where the cylinders cross spheres and cones

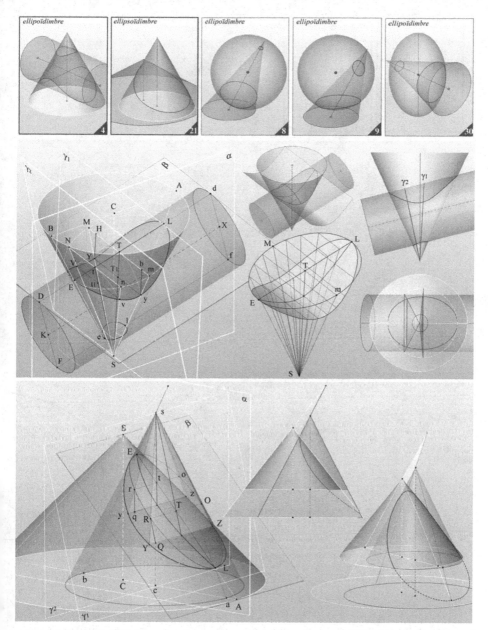

Fig. 19. Examples of intersection between quadrics, where the curve of the fourth order is an *ellipsoïdimbre*, specifically, reconstruction of the examples of intersection where the cones cross cylinders and other cones

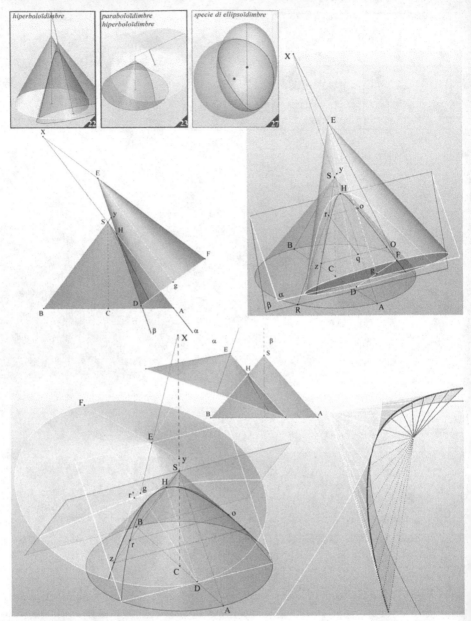

Fig. 20. Examples of intersection between quadrics, where the curve of the fourth order is a *paraboloïdimbre* and an *hyperboloïdimbre*, specifically, reconstruction of the examples of intersection regarding two cones

## Acknowledgments

Where not otherwise indicated, all graphic eleborations are by the author.

# Notes

1. In a rather hasty and somewhat daring manner, the paternity of descriptive geometry was often attributed to Gaspard Monge, but like the genesis of all sciences, in the case of descriptive geometry it is short-sighted to ignore the theoretical and practical contributions that contributed to creating it.

2. By "associated projections" I refer to the graphic method of double orthogonal projection. Since it was Gaspard Monge who invented the term and classified the method, I have preferred to use the definition used before the method theorised by Monge.

3. There is a close link between improvements in the graphic method and the evolution of stonecutting techniques. For example, the realistic feasibility of certain works such as the spiral staircase known as 'the screw of Saint Gilles', in which the edges of the ashlars belong respectively to the curves of double curvature and to skew surfaces, prove that pre-Renaissance building techniques needed drawings and signs to be built.

4. The French word *trait* (pl. *traits*) in stereotomy defines drawings (in scale or proportion) that generally consist of a plan, of one or more rotated sections, and one or more developed plans. This kind of drawing contains the geometrical and constructive information need to carry out a specific work of cutting stone.

5. Poudra writes that only in 1822 Poncelet in his *Traité des proprietés projectives* drew people's attention to this important geometrician, ennobling his name as one of the founders of modern projective geometry. The same holds true for Michel Chasles, who in *Aperçu Historique sur l'origine et le développement des methods en Géométrie* [1837] published his studies on Desargues's work and disseminated several new geometric concepts. In a bookshop in Paris Chasles also discovered a book transcribed by Philippe de La Hire of one of the most important works by Desargues: *Brouillon project d'une atteinte aux événements de la rencontre d'un cone avec un plan*, which was published, after its discovery, by Noël-Germinal Poudra in *Oeuvres de Desargues reunies et analysées par M. Poudra*, in 1861.

6. Even if Hachette recognised the good work done by Frézier, he ended his report by saying that Monge was the one who gave a new dimension to the entire field of the science of projections [Hachette 1816].

7. Stonecutters knew how to create the edges of double curvature of an ashlar physically, but there was no theory to support their expert knowledge and, therefore, no knowledge of the geometric nature of the curves they used.

8. Frézier also makes reference to the treatise by the Jesuit father Pierre Coucier, *De Sectione Superficiei sphaericae, clindricae par cylindricam et conicae per conicam di Pierre Courcier* (1663), which explains the geometric nature of the curves of double curvature created by the intersection of elementary solids. Even if Courcier's contribution presents no particular difficulties as far as the solution of the problems of intersection is concerned, it methodologically anticipates the analysis proposed by Frézier in his *Traité de stéréotomie*.

9. As early as 1530 Nonius, and a little later Wright, Stevin and Snellius studied rhumb lines: Halley considered this curve as the stereographic projection of a logarithmic spiral. Then came the studies by Roberval published in 1630 in the *Treatise of the Indivisibles* dedicated to the curve which Loubère later defined as *cyclocylindrical*. This was followed by the studies by Descartes which, at the end of the second book of his *Géométrie* in 1637, introduced the revolutionary concept of surfaces of double curvature in the doctrine of coordinates in space. The first to represent a curved surface using an equation with three variables was Antoine Parent (1666-1716). Then came several studies dedicated to the solution of single problems that fascinated mathematicians between the late seventeenth and early eighteenth century, including those by Johann Bernoulli and Leonhard Euler. However it was Clairaut who wrote a comprehensive and complete treatise on curves of double curvature (1713-1765) [See Chasles 1837: 136-141].

10. In this work Clairaut analyses geometric and not transcendent curves. However, as he himself writes in the preface, the principles described can be applied to this kind of curve even if special methods have to be used.

11. The work is divided into four sections: how to consider curves of double curvature; the use of

differential calculus compared to their tangents and perpendiculars; the use of integral calculus compared to their rectification; the quadrature of the spaces they create and, finally the illustration of certain general principles required to build curves of double curvature and understand their genesis.

12. For biographies of Frézier, see [Desessarts 1800: 175-177; Faller 1792: 214].

13. Amédée François Frézier (1682-1773) is known as a great military strategist, an expert in fortifications and an enthusiastic traveller. He spent his entire career as an engineer and member of the Army Intelligence Corps, a job which gave him the possibility to travel extensively. He wrote many books on military architecture as well as works which span many other fields of learning, including *Traité des feux d'artifice* and *Relation du voyage de la mer du Sud aux côtes du Chili, du Pérout et de Brésil, fait pendant les années 1712, 1713, 1714. Ouvrage enrichi de quantité de planches en taille-douce*, which includes detailed maps of the east coast of Latin America. His travels and inclination for theory and study, coupled with his work which made him a protagonist of eighteenth-century fortifications, gave him extensive knowledge in many fields of learning. He spent most of his life on military worksites, where he was able to experiment with different building techniques and at the same time improve his skills in the art of war.

14. The definition "scientific stereotomy" was coined by Loria [1921: 87-96].

15. Even if the entire treatise is dedicated to engineers and architects, Frézier explicitly dedicates the first volume to the Learned, who he defines as "those who focus on difficult things", in other words to the men of science who from this one theory will be able to independently elaborate various applications, if they are curious enough.

16. Stone masonry architecture consists of ashlars that are dressed elsewhere based on principles of prefabrication. When assembled their geometry creates the overall object, without the use of mortar or concrete. When used, mortar had no static function, but was used to fill in any unevenness and surface imperfection between each ashlar, making them perfectly flat so they could evenly distribute the pressure exerted by one element on another and prevent any cracking that might be caused by bad alignment [Salvatore 2009: 489-491].

17. Frézier cites several authors in particular but not their works. It is reasonable to assume he was referring to: Philippe de La Hire (1640-1718) for his *Traité de mécanique, où l'on explique tout ce qui est necessaire dans la pratique des Arts* (1695) and *Sur la construction des voûtes dans les édifices* (1712); Claude Antoine Couplet (1642-1722) for his *De la pousée des voûtes* (1729); Antoine Parent (1666-1716), for his *Élements de méchanique et de phisique: où l'on donne géometriquement des principes du choc et des équilibres entre toutes sortes de corps; avec l'explication naturelle des machines fondamentales* (1700), *Nouvelle statique avec frottemens et sans frottemens ou Regle pour calculer le frottemens des machines dans l'état d'équilibre* (1704), *Des résistance des tujaux cilindriques pour des charges d'eau, & des diametres données* (1707), *De résistance des poûtres par rapport à leur longeurs ou portées, & à leur dimensions et situations; et des poûtres de plus grande résistance, indépendamment de tout système physique* (1708) and *Des points de rupture des figures: de la manière de les rappeler à leurs tangentes: D'en déduire celles qui son partout par-tout d'une résistance égale: Avec la méthode pour trouver tant de ces sortes de figures qui l'on veut: Et de faire en sorte que tout sorte de figure soit partout d'une égale résistance, ou ait un ou plusieurs points de rupture* (1710); Bernard Forest de Belidor (1697-1761), for his *La science des ingégneurs dans la condite des travaux de fortification et d'architecture civile* (1729). Most of these works were published for the first time in *Mémoire de l'Académie Royale des Sciences*.

18. As mentioned earlier, the treatises on steretomy published before Frézier were in fact constructive algorithms rarely accompanied by demonstrations, except in the work by the Jesuit Father Claude François Milliet Deschales entitled *De Lapidum Sectione* (a chapter of the mathematical work *Cursus seu Mundus mathematicus* published in 1674).

19. This is the definition given by the author in his treatise.

20. The terminology used by Frézier to describe quadric surfaces and curves of the fourth order is different from the one we use today. In the treatise the author makes extensive use of the term "bodies," meaning volumes containing quadric surfaces which, depending on their type, are called conic surfaces, cylindrical surfaces etc. and are divided by a plane surface that has any

conic section as its perimeter. Instead he uses Clairaut's definition for space curves, i.e., curves of double curvature.

21. Frézier considers 'regularly irregular' bodies to be all kinds of round bodies, even if he includes helicoids.

22. A *Spheroid* is a surface that we today would call an *ellipsoid of rotation*; this term was used by ancient geometricians and is present in the *De conoidibus et sphoeroidibus,* in which Archimedes introduces three solids: the *hyperbolic conoid,* the *parabolic conoid* and *elongated* or *prolate spheroids.*

23. Frézier refers the reader to the study of *Conics* by Apollonius, the *Les éléments des section coniques* by Philippe de La Hire (1679) and the *Traité analytique des sections coniques* by the Marquis de l'Hôpital (published posthumously in 1707).

24. In particular, the *Brouillon projet d'une atteinet aux événements des rencontres du cône avec un plan,* written by Desargues in 1639, the *Essay sur les coniques* written by Pascal in 1640 and the *Sectiones conicae in novem libros* written by de La Hire in 1685, but also the work by Guarini entitled *Trattato sulle coniche* (1671) or the work by Le Piovre entitled *Traité des sections du cylindre et du cône, considerées dans le solide et dans le plan, avec demonstrations simplex et nouvelles* (1704).

25. It would be easy to illustrate the immense resources that this method would have immediately afforded geometricians if the latter had not abandoned the lessons of old geometry [see Chasles 1837: 116-141].

26. Frézier developed this concept in *Élements de stéréotomie à l'usage de l'architecture* [1760], which summarises the contents of the *Traité de stéréotomie.*

27. In *Élements de stéréotomie à l'usage de l'architecture* [1760], Frézier only includes the solution proposed for the cylinders, but not the cones.

28. By "mathematical modelling" we mean three-dimensional modelling using Nurbs curves and surfaces to show forms in space, so curves and surface are described using a parametrical equation.

29. If the two surfaces were not equidistant, and the cylinder was created by tracing two straight lines around two concentric ellipses, the positions of the circular sections would lie on different planes and would give rise to two special cases, one in which the section of the external cylinder is a circular circumference and one where the cylinder is an ellipse, and vice versa.

30. These words come from the Latin *Imbrex,* "hollow tile," which Frézier believed described the form of the sections created by intersection of the bodies. These are neologisms which appear in his treatise for the first time, but which did not become popular.

31. This is the abbreviation of the Latin expression, *ellipsis imbricata.*

32. Frézier explains that the name comes from the fact that this curve somehow imitates the *ellipsimbre.*

33. With reference to the book to be bound, the number of pages that slide when the spine is rolled remains the same; since the pages can be compared to the ordinates of a conic, one can conclude that during projection the latter remain the same in number and, only in the case of the *cicloïmbre* and *ellipsimbre,* also in size.

34. Frézier uses the term 'solid section' to indicate a portion of a quartic surface defined by a curve of double curvature.

35. The enunciation of this theory is published in [Frere Gabriel Marie 1887/1996: 645-646]. Consideration should be given to the note at point 1053 by M. E. Lemoine.

## References

CHASLES, Michel. 1837. *Aperçu Historique sur l'origine et le développement des méthodes en Géométrie.* Bruxelles: M. Hayez, Imprimeur de l'Académie Royale.

CLAIRAUT, Alexis Claude. 1731. *La Recherche sur les courbes à double courbure.* Paris: Nyon, Didot & Quillau.

COUTRAULT, Luois, ed. 1903. *Opuscules et fragments inédits de Liebnitz.* Paris: Presses Universitaires.

DE LA RUE, Jean Baptiste. 1728. *Traité de la coupe des pierres.* Paris: Imprimerie Royale.

DE L'ORME, Philibert. 1567. *Le premier tome de l'architecture.* Paris: editions Federic Morel.

DERAND, François. 1643. *L'Architecture des voutes Ou l'Art Des Traits, Et coupes Des Voutes ....* Paris: S. Cramoisy.

DESESSARTS, N.L.M., et al. 1800. *Les siècles littéraire de la France, ou nouveau dictionnaire historique, critique et bibliographique de tous les Écrivains français, morts et vivants, jusqu'à la fin du XVIIIe siècle,* volume III. Paris: Imprimeur librairie Place de l'Odeon, Paris.

EVANS, Robin. 1995. *The projective cast. Architecture and Its Three Geometries.* Cambridge: the MIT Press.

FALLER, François Xavier. 1792. *Dictionnaire historique; ou histoire abregée des hommes qui se sont fait un nom par le génie, les talents, les vertus, les erreurs etc.,* volume IV. Ausbourg: éditions M. Rieger.

FRERE GABRIEL MARIE (F.G.M. - Edmond Brunhes). 1877. *Géométrie descriptive - Éléments.* (anastatic reproduction of the 5th ed. (1893) Mayenne: Jacques Gabay, 1996).

———. 1887. *Géométrie descriptive - Exercices.* (anastatic reproduction of the 5th ed. (1920) Mayenne: Jacques Gabay,1996).

FREZIER, Amédée François. 1737-1739. *La théorie et la pratique de la coupe de pierres et de bois pour la construction des voûtes, et autre parties des bâtiments civils et militaire, ou traité de stéréotomie à l'usage de l'architecture.* Strasbourg-Paris: Jean Daniel Doulsseker-L. H. Guerin.

———. 1760. *Élémens de stéréotomie à l'usage de l'architecture pour la coupe des pierres.* Paris: Jombert.

HACHETTE, Jean Nicolas Pierre. 1816. Rapport fait à la classe des Sciences Physique et Mathématiques de l'Institut... Pp. 234-237 in *Correspondance sur l' École Royale Polytechnique, à l'usage des élèves de cette École,* volume III. Paris: Imprimerie de M.me V. Courcier.

KLINE, Morris. 1991. *Storia del pensiero matematico, dall'antichità al settecento,* volume I. Torino: Einaudi.

LORIA, Gino. 1921. *Storia della geometria descrittiva dalle origini ai giorni nostri.* Milano: Hoepli.

———. 1931. *Il passato e il presente delle principali teorie geometriche,* IV ed. Padova: Cedam edizioni.

LAURENT, Roger and Joël SAKAROVITCH. 1989. Il trattato del taglio delle pietre di Girard Desargues. Pp. 84-87 in *Atti del convegno, I fondamenti scientifici della rappresentazione,* (Università La Sapienza di Roma, Dipartimento di Rappresentazione e Rilievo, Roma), Cristiana Bedoni and Luigi Corvaja, eds. Rome: Kappa.

MERSENNE, Marin. 1644. *Universae geometriae mixtaeque mathematicae synopsis.* Paris: Bertier.

MILLIET DECHALES, Claude François. 1690. De *Lapidum Sectione in Cursus seu mundus mathematicus ....* Lugduni (Lyon): Anissonios Joan Posuel & Claud Rigaud.

MONGE, Gaspard. 1795. *Géométrie Descriptive* (anastatic reproduction 1989. Sceaux: Jacques Gabay).

OLIVIER, Théodore. 1852. *Cours de Géométrie descriptive, première partie. Du point, de la droite et du plan.* Paris, Carilian-Goelry V^or Dalmont.

PEIFFER, Jeanne. 1995. *Géométrie.* Paris: Seuil.

POTIÉ, Philippe. 1996. *Philibert De l'Orme, figures de la pensée constructives.* Marseille: Editions Parenthèses.

POUDRA, Noël Germinal. 1861. *Oeuvres de Desargues reunies et analysées par M. Poudra,* volume I. Paris: Leiber Éditeur.

SAKAROVITCH, Joël. 1998. *Épures d'architecture: de la coupe des pierres à la géométrie descriptive XVIe XIXe siècle.* Basel-Boston-Berlin: Birkhäuser Verlag.

SALVATORE, Marta. 2008. La "stereotomia scientifica" in Amedée François Frézier. Prodromi della geometria descrittiva nella scienza del taglio delle pietre. Ph.D. dissertation, Florence.

———. 2009. La stereotomia. Pp. 485-523 in Riccardo Migliari, *Geometria Descrittiva,* vol. II. Novara: CittàStudi Edizioni.

TATON, René. 1951. *L'oeuvre mathématique de Girard Desargues.* Paris: Presse universitaire de France.

VANDELVIRA, Alonso. 1580 c. *Libro de trazas de corte de piedra* (anastatic reproduction *Tratado de arquitectura de Alonso de Vandelvira,* Genèvieve-Barbé Coquelin de Lisle, ed. Albacerte: Caja de Ahorros, 1997).

---

## About the author

Marta Salvatore, architect and Ph.D. in Survey and Representation of Architecture and the Environment is currently holds a research grant in the Department of History, Design and Restoration of Architecture at "La Sapienza" University of Rome. Her research activity is directed towards the study of Descriptive Geometry and the disciplines at its base. Research of the theory in the ancient texts, attention to the evolution of graphic methods of the representation and transposition in space of problems of classic descriptive geometry constitute the underlying theme that animates her research activity. In this context she deals with the problem of renewing descriptive geometry through information technology methods of representation, making it possible to surpass the limits imposed by graphic methods, thus taking another look at descriptive geometry as a continually evolving discipline with an aim towards extending knowledge to heretofore unexplored areas.

# Snežana Lawrence

School of Education
Bath Spa University
Newton Park
Bath, BA2 9BD, United Kingdom
snezana@mathsisgoodforyou.com

Keywords: developable surfaces,
Gaspard Monge, descriptive
geometry, ruled surfaces, Frank
Gehry, Leonhard Euler

Research

# *Developable Surfaces: Their History and Application*

Presented at Nexus 2010: Relationships Between Architecture
and Mathematics, Porto, 13-15 June 2010.

**Abstract.** Developable surfaces form a very small subset of all
possible surfaces and were for centuries studied only in
passing, but the discovery of differential calculus in the
seventeenth century meant that their properties could be
studied in greater depth. Here we show that the generating
principles of developable surfaces were also at the core of their
study by Monge. In a historical context, from the beginning
of the study of developable surfaces, to the contributions
Monge made to the field, it can be seen that the nature of
developable surfaces is closely related to the spatial intuition
and treatment of space as defined by Monge through his
descriptive geometry, which played a major role in developing
an international language of geometrical communication for
architecture and engineering. The use of developable surfaces
in the architecture of Frank Gehry is mentioned, in particular
in relation to his fascination with 'movement' and its role in
architectural design.

## Introduction

Contemporary use of developable surfaces and their geometric manipulation with or
without the use of computer software has been increasingly well documented (see in
particular [Liu et al. 2006] and the bibliography therein). The application of developable
surfaces is wide ranging – from ship-bulding to manufacturing of clothing – as they are
suitable to the modelling of surfaces which can be made out of leather, paper, fibre, and
sheet metal. Some of the most beautiful works of modern architecture by architects such
as Hans Hollein, Frank O. Gehry and Santiago Calatrava use the properties of
developable surfaces, yet the history of this type of surface is not well known; this is the
task the present paper is set to achieve.

Developable surfaces form a very small subset of all possible surfaces: for centuries
cylinders and cones were believed to be the only ones, until studies in the eighteenth
century proved that the tangent surfaces belong to the same mathematical family (see in
particular [Euler 1772] and [Monge 1780, 1785]).

There are therefore three types of developable surfaces (excluding a fourth type, the
planar surface):

1. surfaces in which generating lines are tangents of a space curve: this type of
   surface is spanned by a set of straight lines tangential to a space curve, which is
   called the edge of regression (fig. 1a);
2. surfaces which can be described as a generalised 'cone' where all generating lines
   run through a fixed point, the apex or vertex of the surface (fig. 1b);
3. surfaces which, in the same manner, can be described as a generalised cylinder,
   where all generating lines are parallel, swept by a set of mutually parallel lines
   (fig. 1c).

DOI 10.1007/s00004-011-0087-z; *published online* 22 November 2011

It should be mentioned that the developable surfaces are also those which contain elements of any of the above mentioned general cases of developable surfaces, as long as they can be flattened onto a planar surface, without creasing, tearing or stretching (fig. 1).

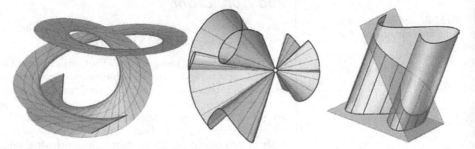

Fig. 1. The three kinds of developable surfaces: a, left) tangential; b, centre) conical; c, right) cylindrical. Curves in bold are directrix or base cruves; straight lines in bold are directors or generating lines (curves)

Developable surfaces are a special kind of ruled surfaces: they have a Gaussian curvature equal to 0,[1] and can be mapped onto the plane surface without distortion of curves: any curve from such a surface drawn onto the flat plane remains the same. In this context, it is important to remember the following property of Gaussian curvature: if the surface is subjected to an isometric transformation (or more plainly bending), the Gaussian curvature at any point of the surface will remain invariant.[2] The Gaussian curvature is in fact determined by the inner metric of a surface, therefore all the lengths and angles on the surface remain invariant under bending, a property immensely important in using developable surfaces in manufacturing. As a consequence, developable surfaces, having the Gaussian curvature equal to zero, the same as plane surfaces, can be obtained from unstretchable materials without fear of extending or tearing, but by transforming a plane through folding, bending, gluing or rolling.

The use of developable surfaces in contemporary architecture, for example in the work of Gehry,[3] to which we will return later on in this paper, have been, more recently, made possible by the development of computer software which, given user-specified three-dimensional boundary curves, generates a smooth developable surface.

In our usual world of three dimensions, the one in which we make architectural objects, all developable surfaces are ruled.

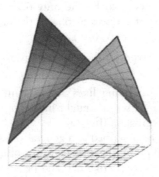

Fig. 2. A hyperbolic paraboloid is a ruled, but not developable, surface

The ruled surface is such that it contains at least one family of straight lines. Ruled surfaces are generated by a *directrix* or *base curve* – the curve along which the ruling, director, or generating curve moves. The *rulings* or *directors* of ruled surfaces are straight lines. Although all developable surfaces (in three dimensions) are ruled, not all ruled surfaces are developable. For example, the hyperbolic paraboloid is not developable but it is a ruled surface (fig. 2).

To summarise: the developable surfaces are cylinders, cones, and tangent developable surfaces, or a composition of these. The tangential developable surfaces can be best described and visualised as surfaces formed by moving a straight line in space (the director as explained above) along a directrix. If you imagine a generating line describing a tangential surface, then all points on that generating line share a common tangent plane.

## The history of developable surfaces

The history of developable surfaces can be traced as far back as Aristotle (384-322 B.C.): in *De Anima* (I, 4) he states that 'a line by its motion produces a surface' [Aristotle 2004: 409], although this does not mean that it was Aristotle who defined or named the ruled or developable surfaces. Nevertheless, the statement had profound influence on perceiving the generation of surfaces through movement. More than twenty centuries later, Monge (1746-1818) used the principle of generating surfaces in a task he was given while he worked at the Mézières[4] as a draftsmen in the 1760s [Taton 1951; Sakarovitch 1989, 1995, 1997], which led to his discovery of a technique which later gained the name of 'descriptive geometry'. We will return to his invention, but first let us look at how similar treatment of motion was used in practical geometry before him.[5]

A first example is William Hawney, a minor eighteenth-century author on surveying, who in 1717 described the cylinder as a surface 'rolled over a plane so that all its points are brought into coincidence with the plane'.[6] In 1737 Amédée François Frézier (1682-1773) also considered the rolling of the plane to form a circular cylinder and cone (see [Frézier 1737-39]), but he did not generalise on the mathematical properties of this process, nor did he distinguish between developable and ruled surfaces.

Almost half a century later, Euler (1707-1783) and Monge became interested in developable surfaces and used differential calculus to study their properties. Only in 1886 however, was the term "differential geometry" coined. It was used for the first time by Italian mathematician Luigi Bianchi in his textbook *Lezioni di geometria differenziale* (Pisa, 1886). The investigations of Euler and Monge therefore preceded the beginning of a study of differential geometry, and initiated investigations in the field of developable surfaces.

Before we get there, a reminder needs to be made about the nature of the study of change through differential calculus. In the seventeenth century, independently of each other,[7] Isaac Newton (1643-1727) and Gottfried Leibniz (1646-1716) discovered calculus, which deals with the study of change. Newton used calculus to determine an expression for the curvature of plane curves. As the study of tangential surfaces is mentioned throughout this paper, it is worth noting the comparison of the study of curves in two dimensions through differential calculus. For example, finding a tangent to a curve at any point involves seeking the first derivative of that curve and using this result to find an equation of a tangent at a particular point to the curve.

Euler, who was by that time blind, wrote a paper entitled "De solidis quorum superficiem in planum explicare licet" (On solids whose surface can be unfolded onto a plane, E419) [1772] in which his perception of space is clearly shown to be that of identifying and describing surfaces as boundaries of solids, not as collection of solids.[8]

Euler's approach includes differential treatment combined with geometrical. First he looks at any surface and assumes that a limiting value from this surface will be its derivative: this he obtains through selecting an infinitesimally small right triangle on the surface and seeking its relationship to a congruent triangle in a plane. At this point Euler is generalising a problem via a system of differential equations, but he then proceeds to seek its solution via geometrical treatment. Euler draws three lines on a sheet and describes them simply as Aa, Bb, Cc (fig. 3).

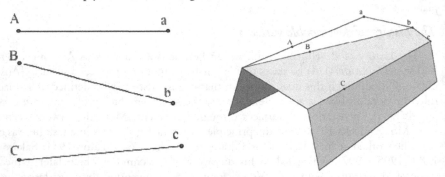

Fig. 3. Euler's generalisation begins from three straight lines on a sheet of paper, and the bending of the paper along those lines

He then proceeds to investigate what happens when this sheet is bent along a straight line; however that happens, it is always possible to conceive of a solid which fits that bent sheet. From this it follows, Euler concludes that, besides prismatic and pyramidal bodies, there are any number of other kinds of bodies which may be covered in this manner by such a sheet, and whose surfaces may accordingly be unfolded upon a plane:

> Let us now increase to infinity those lines Aa, Bb, Cc, etc. so that our solid acquires a surface everywhere curved, as our problem demands according to the law of continuity. And now it appears at once, that the surface of such bodies should be so constituted that from any point in it at least one straight line may be drawn which lies wholly on this surface; although this condition alone does not exhaust the requirements of our problem, for it is necessary also that any two proximate straight lines lie in the same plane and therefore meet unless they are parallel (translation by Florian Cajori [1929]).[9]

The enormity of Euler's contribution in this respect is simple to establish. He extended the study of surfaces to developable surfaces by posing a simple question about *which kind of surfaces can be developed into a plane*:

> *Notissima est proprietas cylindri et coni, qua neorum superficiem in planum explicare licet atque adeo haec proprietas ad omnia corpora cylindrical et conica extenditur, quorum bases figuram habeant quamcunque; contra vero sphaera hac proprietate destituitur, quum eius superficies nullo modo in planum explicari neque superficie plana obduci queat; ex quo nascitur quaestio aeque curiosa ac*

*notatu digna, vtrum praeter conos et cylindros alia quoque corporum genera existant, quorum superficiem itidem in planum explicare liceat nec ne? quam ob rem in hac differtatione sequens considerare constitui Problema:*

Inuenire aequationem generalem pro omnibus solidis, quorum superficiem in planum explicare licet, *cuius solutionem variis modis sum agressurus.*

[Euler 1772: 3].

Euler proceeds to look at the intersections of the pairs of lines: if connected these intersections themselves would form a twisted surface of double curvature, a surface, as it were, of a 'higher' degree than the one from which we started (of single curvature, a developable surface). Euler then established the analytical relationship between this twisted curve and the points on the developable surface. His exposition was based on three aspects of the study of such surfaces: the study of developable surfaces by the means of analytical principles; their study by the means of geometrical principles; and finally the study in which the second is applied to the study of the first. It is interesting to note that, although Euler's opus includes some 832 original works,[10] nevertheless this paper of Eulier is considered by some to be his very best mathematical piece.[11]

## Monge's study of developable surfaces and the invention of Descriptive Geometry

Independently from Euler, and at almost the same time, a young draughtsman glimpsed a new way of imagining spatial relations through principles of generation of geometrical entities by movement. Monge therefore studied developable surfaces, having started from an entirely different viewpoint. In, or around, 1764, and while working at the l'École Royale du Génie de Mézières, Monge was given the task of determining the necessary height of an outer wall in a design of fortification (fig. 4).

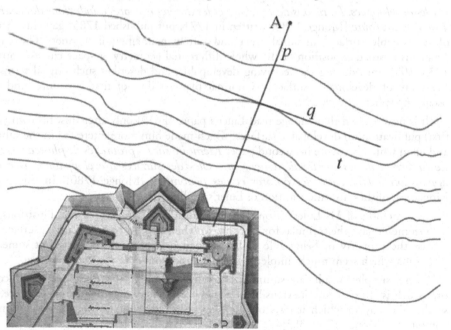

Fig. 4. Monge's work on determining the height of the fortification wall included construction of developable surfaces

Up to that time, there were two methods most widely used for this problem. One involved determining numerous view points from the terrain; the triangles thus determined by the view point, a point of the edge of the fortification, and the height of the wall sufficient to offer effective protection, were all necessary measurements; a lengthy empirical process was involved. The other method, which was employed in the school in Mézières, was based on long calculations with the heights of each crucial point being measured directly on the terrain and noted on a plan. Instead of using these methods, which would have taken a week to yield the final result, Monge found a way to finish his work in two days. Monge conceived of a new technique, which he called *Géometrie Descriptive* and which was naturally related to the methods widely used at the time. The method was examined with the highly critical eye of his superintendents who had their suspicions about Monge's speed in resolving the problem; when properly understood, it was ruled a military secret.

To solve the problem he was faced with, Monge determined a plane tangent to the terrain. This plane is determined by a point and a line: point $A$ lies in the ground plane of the plan of fortification (fortification designs were usually predetermined and their plans were strictly geometrical), line $q$ is drawn as a perpendicular to line $p$ ($p$ is a tangent to $t$), which is the contour line of the terrain).

Monge further considered conical surfaces which used lines such as $p$ as generatrixes and points such as $A$ as centres to determine the height of the fortification (the thickened line perpendicular to the plane in which the fortification plan rests).

Once it became a military secret, Descriptive Geometry saw little light until the reform of the educational system of France, which took place during the Revolution. However, Monge's first publications were on developable surfaces: these came years before his technique of descriptive geometry could be made public. His paper, *Mémoire sur les développées, les rayons de courbure et les différents genres d'inflexions des courbes à double courbure* [Monge 1785; written in 1771, but published 1785] gave his theory of developable surfaces in an abstract and purely mathematical manner. The paper "contains a broad exposition of the whole differential geometry of space curves" [Struik 1933: 105], introducing the rectifying developable, and describes such crucial terms in the study of developable surfaces as normal plane, radius of first curvature, and the osculating sphere [Reich 2007].

It is now known that Monge read Euler's paper (E419) only after this, his own (and first) publication on developable surfaces, which made him more interested in the subject and upon which he wrote his second paper *Mémoire sur les propriétés de plusieurs genres de surfaces courbes, particulièrtement sur celles des surfaces d'eveloppables, avec une application à la theorie des ombres et des pénombres* [Monge 1780]. In this paper Monge establishes his simplification of Euler's findings

> ... a memoir of Mr Euler ... on developable surfaces ... in which that illustrious Geometer gave the formulas for recognizing whether or not a given curved surface has the property of being able to be mapped to a plane, ... I arrived at some results which seem much simpler to me and easier to use for the same purpose.[12]

These 'simpler' results are summarized in his definition of a developable surface as one which is "flexible and inextensible, one may conceive of mapping it onto a plane, ... so that the way in which it rests on the plane is without duplication or disruption of continuity" [Monge 1780: 383].

In this, his second paper on the topic, Monge begins with a curve of double curvature (a twisted curve – for example, a helix) and defining a point on it. Through this point he draws a plane which is perpendicular to a line tangent to the curve at this point; he does the same with another point, tangent and plane that go perpendicularly through this tangent. The two planes intersect in a line. If one imagines a curve which goes around the original surface (of double curvature), takes consecutive points on it and does the same as above (i.e., draws through each point a tangent, and through each tangent a normal plane), then finds the intersections of these planes, a developable surface will have thus been constructed. The generalisation to which Monge arrived at was that, in fact, any curvature of double surface can be enveloped by a space curve; this tangent curve can be used, as explained, to generate a developable surface. Monge's spatial and Euler's analytical insights, proved that the tangent surfaces belong to the same family as conical and cylindrical ones.

Further, developable surfaces are, as will be shown, an integral part of the Mongean treatment of space. His descriptive geometry was born out of an insight into a possibility of constructing an imaginary tangential plane to a terrain in order to solve the real problem of fortification design. But it was not only that his description of the technique itself is a study in developable surfaces; it was that in fact, all objects imagined through the use of descriptive geometry as given by Monge are ruled surfaces (and most are developable). Monge describes all geometrical objects through generation: a point is a generatrix of a line; similarly any plane is generated by two lines. In descriptive geometry the position of any element is determined by its position with respect to the projection planes, of which there are two (for simplicity of explanation, the horizontal and vertical). The generation of a plane surface could be described by the lines in which the plane in question intersects the projection planes. These two lines determine the plane in full and are called the traces of the plane.

In order to arrive at a fuller understanding of Monge's conception of space we will bring one example as follows. The traces of the plane on the projection planes may also be regarded as the lines that generate the plane. It should be now easy to see that those two traces meet on the straight line in which the planes of projection meet each other. In Monge's drawing (fig. 5) this means that in the plane, which has been named from its traces as BAb, those traces, namely the line AB and Ab, meet on the line AC=LM, which represents the intersecting line of two projection planes.

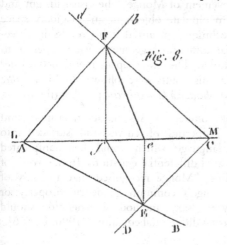

Fig. 5. Plate III from the 1811 Hachette édition of Monge's *Géométrie Descriptive* showing the primary operations with two planes. The system, although determined by two planes of projection, is actually represented only by their line of intersections, AC=LM on this diagram

If we were to consider a second plane (this plane is labelled DCd) we would be able to see how this plane meets the first plane, and their intersection line is EF, the first projection of which is given as Ef and second as eF. We may bear in mind that this is the process that is used in the construction of any solid body with plane surfaces for its sides.

Two important things that explain, in basic terms, the principles of the technique, are to be learnt from this example. First, every three-dimensional geometrical body can be described as a sum of planes which are generated by some mode of motion and intersect each other. This means that the solid is not regarded as an independent entity in itself, but as a product of motion and intersection of its primary elements. Had we further elaborated upon the above example, and said that, for example, the two planes which intersect should be orthogonal to each other, and if we had continued by adding four further planes in a similar manner so that each of them is orthogonal to the one it intersects, and that they should be on the same distance to the parallel ones on all sides, we would generate a cube. In this way the perception of all geometrical entities is changed from a collection of various forms, to a collection of various methods and processes by which those forms are being generated.

Monge realised that, although descriptive geometry had a very practical application in stone-cutting, engineering, and architectural drawing, this treatment of space had also implications for the study of the properties of curves. To Monge, developable surfaces were not just mathematical curiosities which arose out of his spatial insight underpinned by the technique of descriptive geometry which he perfected. In fact, the study of surfaces was an integral part of his interest in stonecutting; in the first edition of *Descriptive Geometry* he describes, for example, how the surfaces of the voussoir joints must be developable

> ...This is a practical consideration: if such a surface were not a ruled surface, it would not be executable sufficiently rapidly to be economically viable or with sufficient precision to ensure proper contact between the voussoirs. The fact that the surface is also developable allows greater precision with the use of panels for tracing [Monge 1795b, quoted in Sakarovitch 2009: 1295].

Fig. 6. One of Monge's main works on developable surfaces, *Feuille d'Analyse* [1795a]

But being practical, however useful, was not the main achievement of Monge. The visual insight and ability to manipulate objects in an imaginary space with his technique, meant that his results in three-dimensional differential geometry were superior to those of Euler. The two papers mentioned earlier summarising his results were followed by his *Feuilles d'analyse appliquée à la géométrie* [1795a] (fig. 6).

This book offered a systematic treatment and translated properties of curves and surfaces into analytical language which could be manipulated through partial differential equations. In the words of Morris Kline, "Monge recognised that a family of surfaces having a common geometric property or derived by the same method of generation should satisfy a partial differential equation" [1990: II, 566].

Monge's treatment of developable surfaces is, as can be glimpsed from these short insights into his work on the topic, entirely different from that of Euler. While Euler's treatment was 'of a profoundly analytic spirit' [Taton 1951: 21], Monge had keen geometrical intuition which manifested itself most profoundly in his conception of descriptive geometry, and which enabled him to apply analysis to geometry, rather than consider the two disciplines as separate ways of inquiry.

## The collaborative fraternity of Monge and Gehry

Separated by centuries and different professions, it may seem that Monge and Gehry may not have anything in common apart from an interest in developable surfaces. And perhaps one would be right to disregard any further analysis of their common interest. Let us however, entertain an idea that both, in their own times and in their own ways, were interested in these surfaces not by chance, but because they embodied certain ideas about motion that had a particular slant on creative processes that resonated with both Monge and Gehry. In Monge's case this manifested itself in his involvement in all social, political, and educational aspects of his involvement in building the new Republic. Monge built on the tradition and knowledge of stonemasons, resulting in his conception of a new, all-encompassing technique which would serve as a language of graphical communication throughout the territory in which French educational system would extend its influence. This proved to be a large territory, with influences felt up to modern times, and with descriptive geometry still surviving in many national educational systems.[13] Monge's wholehearted involvement with the ideals of the Enlightenment and the founding of the revolutionary educational institutions of the new Republic, gained him the title of the Father of École Polytechnique (see [Sakarovitch 2009]).

The teaching of descriptive geometry in this institution was certainly not a continuation of the educational tradition of the Ancien Régime; it was a revolutionary subject taught in a revolutionary way, in the first of revolutionary schools:

> A scholastic discipline which was born in a school, by a school and for a school (but maybe one should say in the École Polytechnique, by the École Polytechnique, and for the École Polytechnique), descriptive geometry allows the passage from one process of training by apprenticeship in little groups which was characteristic of the schools of the Ancien Régime, to an education in amphitheatres, with lectures, and practical exercises, which are no longer addressed to 20 students, but to 400 students. Descriptive geometry also stems from revolutionary methods. A means to teach space in an accelerated way in relation to the former way of teaching stereotomy, an abstract language, minimal, rapid in the order of stenography, descriptive geometry permits a response to the urgent situation as for the education of an elite, which was the case of France at the moment of the creation of École Polytechnique [Sakarovitch 1995: 211].

It is here, in the context of such sweeping changes, that we move our focus back to the architecture of Gehry (fig. 7). In particular, his architecture is habitually defined by the use of developable surfaces constructed from sheet metal. Gehry himself lists several things that led him to the use of developable surfaces in his later designs, the most important of which he lists as collaboration, movement, and context. All of these in their own way correspond to some of the sentiments that descriptive geometry inspires by its method. Movement is, for Gehry, the leading force:

Fig. 7. Frank O. Gehry, the Guggenheim Museum Bilbao, completed 1997.
Photo: iStockphoto

I became fascinated with the idea of building in a sense of movement with static materials. Historically there are a lot of references. At the Parthenon, Phidias played with that idea – not in the building, but in the sculptures. And the Indian Siva figures are extraordinary, if you look at the best ones: when you look away, you are sure they moved. That has always fascinated me [Gehry 1996: 38].

His interest in collaboration – the building on the tradition of unions – is somewhat similar to Monge's sentiment regarding the stonemasons;[14] in Gehry's case this manifested itself in his involvement with the metal workers' union. He highly values their input – and testifies to a bond between the designer and the executor in terms of their regard for the material to which both are dedicated and from which new structures are built:

The [Sheet Metal Workers Union of America] is connected to metal workers around the world, and they have agreed to provide me with technical assistance wherever I go. They have lived up to that agreement, most recently on a project in Bilbao, Spain. They have helped me work within my budget and still achieve the forms I want. This collaboration has certainly made possible some of my successes in the use of metal materials [Gehry 1996: 40].

Finally it is movement, in all its symbolic and contextual form, that Gehry perhaps embodies best through his architecture: the sweeping movements of lines generating developable surfaces 'that is the sort of thing that occurs in a fast-moving city; as one tries to respond to the context, it changes, always more rapidly' [Gehry 1996: 42].

## Conclusion

In the nineteenth century the mathematics of developable surfaces temporarily left the realm of the real world. This happened for very many reasons but suffice it to say that, for example, the freer and more common use of the imaginaries of all kinds: talking about 'circular points of the line at infinity' and being able not only to describe but manipulate imaginary numbers with increasing ease, meant that the study of surfaces increasingly involved interest in the imaginaries linked to such entities. This in turn led to the discovery and study of minimal surfaces.[15] In another development, Henry Lebesgue (1875-1941), best known for his theory of integration, made an astonishing discovery of surfaces which, although resembling developable surfaces, are neither developable nor ruled, and can be flattened into a plane without stretching or tearing (but with creasing) [Picard 1922].[16] Lebesgue's own description of such surfaces is somewhat convoluted, but nevertheless its very short version is given in a footnote to a simple explanation given to Lebesgue's discovery by Émile Picard (1856-1941):

> According to general practice, we suppose in the preceding analysis, as in all infinitesimal geometry of curves and surfaces, the existence of derivatives which we need in the calculus. It may seem premature to entertain a theory of surfaces in which one does not make such hypotheses. However, a curious result has been pointed out by Mr Lebesgue (*Comptes Rendus*, 1899 and *thesis*); according to which one may, by the aid of continuous functions, obtain surfaces corresponding to a plane, of such sort that every rectifiable line of the plane has a corresponding rectifiable line of the same length of the surface, nevertheless the surfaces obtained are no longer ruled. If one takes a sheet of paper, and crumples it by hand, one obtains a surface applicable to the plane and made up of a finite number of pieces of developable surfaces, joined two and two by lines, along which they form a certain angle. If one imagines that the pieces become infinitely small, the crumpling being pushed everywhere to the limit, one may arrive at the conception of surfaces applicable to the plane and yet not developable [in the sense there is no envelope of a family of planes of one parameter] and not ruled [Picard 1922], quoted in [Cajori 1921: 436].

An apparent departure from the 'real world' into the abstract construct of imaginaries permitted a new understanding of developable surfaces. But their physical embodiment in the architecture of Frank Gehry perhaps provides the most fitting monument to the study of developable surfaces in three dimensions, and to Mongean treament of space.

## Notes

1. Gausian curvature is equal to zero when the product of minimal and maximal curvatures of a curve is equal to zero, i.e., $k = k_{min} \times k_{max}$. Carl Friedrich Gauss (1777-1855), called sometimes "the Prince of Mathematicians" contributed to a huge array of fields in the study of mathematics, including analysis, differential geometry, number theory, astronomy and optics.
2. Beyond the scope of this paper, this property of Gaussian curvature is none the less fascinating. It was expounded upon in Gauss's *Theorema Egregium* (Remarkable Theorem), which states that the Gaussian curvature of a surface can be determined by measuring distances and angles of a surface without reference to the way in which the surface is positioned in three-dimensional space.
3. For further details see [Amar and Pomeau 1997].

4. The Royal School of Engineering at Mézières was founded in 1748 and was closed in 1794 when it transferred to the School of Engineering at Metz.

5. Omitted from this paper for the reasons of length and simplicity are descriptions of the work of Jean Baptiste Marie Charles Meusiner (1754-1793) and Piere Charles François Dupin (1674-1873). Dupin's main contribution was the invention of a method for describing the local shape of a surface by, what is now called *Dupin indicatrix*. The indicatrix is the end product of a limiting process by which the plane approaches a tangent plane of the surface studied. Meusiner, on the other hand studied minimal surfaces.

6. See [Cajori 1929: 431]; the description used in Hawney is 'in form of a Rolling stone used in Gardens' [1717: 154].

7. Although this fact was vehemently denied by Newton who pursued the claims of plagiarism until Leibniz's death; see [Hall 1980].

8. In fact this approach was followed as late as the nineteenth century in relation to constructive geometry: for a different treatment of space in particular in context of generating surfaces (Gaspard Monge, France) and manipulating objects (Peter Nicholson, England) see [Lawrence 2002; 2010].

9. Meaning that if there is a surface, there lie in it at least two lines which are parallel [Euler 1772].

10. Euler's papers were catalogued by Gustaf Eneström (hence the E numbers). Most of them can be found at The Euler Archive online: http://www.math.dartmouth.edu/~euler/.

11. According to Andreas Speiser (1885-1970), a Swiss mathematician and philosopher of science, quoted in [Reich 2007: 483, note 7].

12. Full text states: *Ayant repris cette matière, à l'occasion d'un Mémoire que M. Euler a donné dans le Volume de 1771, de l'Académie de Pétersbourg, sur les surfaces développables, et dans lequel cet illustre Géomètre donne des formules pour reconnoître si une surfache courbe proposée, jouit ou non de la propriété de pouvoir être appliquée sur un plan, je suis parvenu à des résultats qui me semblent beaucoup plus simples, et d'un usage bien plus facile pour le même objet* [Monge 1780; quoted in Reich 2007: 490, note 15].

13. Currently a research project by the author, to be presented at the Second International Conference on the History in Mathematics Education, Portugal, October 2011.

14. Without wishing to imply any parallels with Gehry's own sentiments in this regard, it can, however, be noted that the ideals that defined Monge's political and social life were also those of the French Revolution – *liberté, égalité, fraternité* (without the terror) – and which were also manifest in Monge's admission to the French Freemasonic order, the *Grand Orient*; see [Lawrence 2002: 118].

15. As already mentioned, Meusiner's study of surfaces via a calculus of variations led him to discover that a surface is minimal if and only if its mean curvature vanishes. The most common 'real-life' explanation of minimal surfaces is their comparison with the soap film formed around a wire boundary.

16. Lebesgue's own explanation of the surfaces that are not ruled yet can be 'flattened' into a plane goes like this:

    To obtain surfaces that are not ruled I take an analytic developable one and upon it an analytic curve C, not geodesic [shortest path between points on a curved surface]. One knows that there exists another developable surface passing through C, such that one can make the two developables applicable to a plane in such a manner that to each point of C, whether considered as belonging to the one or to the other of the two surfaces, there corresponds one and the same point in the plane. The curve C divides the first developable in two pieces, A, B, also the second into two pieces A′, B′.

    Two of the four surfaces (A, A′), (A, B′), (B, B′) are applicable to the plane without tearing or duplication and are indeed such that one can detach from them a finite piece enjoying the above property and containing an arc of C. This C is then a singular line. As before, one may pass from this one singularity to an infinite number of singularities and one thus obtains surfaces applicable to the plane, yet not containing any segment of a straight line [Lebesgue 1899].

### References

ARISTOTLE. 2004. *De Anima*. H. Lawson Tancred, ed. London: Penguin.

AMAR, Ben M. and Y. POMEAU. 1997. Crumpled paper. *Proceedings of the Royal Society* **453**: 719-755.

CAJORI, F. 1929. Generalisations in Geometry as Seen in the History of Developable Surfaces. *The American Mathematical Monthly* **36**, 8: 431-437.

CAYLEY, Arthur. 1850. On the developable surfaces which arise from two surfaces of the second order. *Cambridge and Dublin Mathematical Journal* **V**: 46-57.

————. 1862. On Certain Developable Surfaces. *Proceedings of the Royal Society of London* **12**: 279-280.

CLAIRAUT, Alexis-Claude. 1731. *Recherches sur les courbes à double courbure*. Paris.

COLLIDGE, Julian Lowell. 1940. *A history of geometrical methods*. Oxford: Oxford University Press.

DARBOUX, Gaston. 1904. The Development of Geometrical Methods. *The Mathematical Gazette*, (3:48) 100-106.

DELCOURT, Jean. 2011. Analyse et géométrie, histoire des courbes gauches De Clairaut à Darboux. *Archive of the History of Exact Sciences* **65**: 229-293.

DOMINGUES, João Caramalho. 2008. *Lacroix and the Calculus*. Berlin: Birkhaüser.

DUPIN, Charles. 1819. *Essai historique sur les services et les travaux scientifiques de Gaspard Monge*. Paris: Bachelier.

EULER, Leonhard. 1772. *De solidis quorum superficiem in planum explicare licet* (E419). *Novi commentarii academiae scientiarvm imperiatis Petropolitanae* **XVI** (1771): 3-34. Rpt. *Leonhardi Euleri Opera Omnia*: Series 1, vol. 28, pp. 161-186.

FREZIER, Amédée-François. 1737-1739. *La théorie et la pratique de la coupe des pierres et des bois ... ou traité de stéréotomie a l'usage de l'architecture*. Strasbourg-Paris: Jean Daniel Doulsseker-L. H. Guerin.

GEHRY, Frank O. 1996. Current and Recently Completed Work. *Bulletin of the American Academy of Arts and Sciences* **49**, 5: 36-55.

GLAESER, G. and F. GRUBER. 2007. Developable surfaces in contemporary architecture. *Journal of Mathematics and the Arts* **1**, 1: 59-71.

HALL, A H. 1980. *Philosophers at War: The Quarrel between Newton and Leibniz*. New York: Cambridge University Press.

HAWNEY, W. 1717. *The Complete Measurer*. London.

KLINE, Morris. 1990. *Mathematical Thought from Ancient to Modern Times*, 3 vols. Oxford: Oxford University Press.

LAWRENCE, Snežana. 2002. Geometry of Architecture and Freemasonry in 19th century England. Ph.D Thesis, Open University UK.

————. 2010. Alternative ways of teaching space: a French geometrical technique in nineteenth-century Britain. Pp. 215-224 in *Echanges entre savants français et britanniques depuis le XVIIe siècle*, Robert Fox and Bernard Joly, eds. Oxford: College Publications.

LEBESGUE, H. 1899. Sur quelques surfaces non réglées applicables sur le plan. *Comptes Rendus de l'Académie des Sciences* **CXXVIII**: 1502-1505.

LIU, Y., H. POTTMAN, J. WALLNER, Y-L. YANG, and W. WANG, W. 2006. Geometric Modeling with Conical Meshes and Developable Surfaces. *ACM Transactions on Graphics* **25**, 3: 681-689.

MONGE, Gaspard. 1769. Sur les développées des courbes à double courbure et leurs inflexions. *Journal encyclopédique* (June 1769): 284-287.

————. 1780. Mémoire sur les propriétés de plusieurs genres de surfaces courbes, particulièrtement sur celles des surfaces d'eveloppables, avec une application à la theorie des ombres et des pénombres. *Mémoires de divers sçavans* **9**: 593-624 (written 1775).

————. 1785. Mémoire sur les développées, les rayons de courbure, et les différents genres d'inflexions des courbes à double courbure. *Mémoires de divers sçavans* **10**: 511-50 (written 1771).

————. 1795a. *Feuilles d'analyse appliquée à la géométrie*. Paris. Second ed. with the title *Applications de l'analyse à la géométrie*. Paris : Baudouin, 1811.

————. 1795b. *Géométrie descriptive*. First ed. in *Les Séances des écoles normales recueillies par des sténographes et revues par des professeurs*, Paris. Re-edition pp. 267-459 in *L'Ecole normale de l'an III, Leçons de mathématiques, Laplace, Lagrange, Monge*, Jean Dhombres, ed. Paris: Dunod, 1992.

————. 1850. *Application de l'analyse a la géométrie*, 5th ed. Paris.

PICARD, E. 1922. *Traite d'analyse*, 3rd ed. Paris: Gauthier-Villars et fils.

REICH, Karin. 2007. Euler's Contribution to Differential Geometry and its Reception. Pp. 479-502 in *Leonhard Euler: Life, Work and Legacy*, Robert E. Bradley and C. Edward Sandifer, eds. Amsterdam: Elsevier.

SAKAROVITCH, Joël. 1989. Theorisation d'une pratique, pratique d'une theorie: Des traits de coupe des pierres à la géométrie descriptive. Ph.D. thesis, Ecole d'Architecture de Paris La Villette.

————. 1995. The Teaching of Stereotomy in Engineering Schools in France in the XVIIIth and XIXth centuries: An Application of Geometry, an 'Applied Geometry', or a Construction Technique? Pp. 205-218 in *Between Mechanics and Architecture*, Patricia Radelet-de Grave and Edoardo Benvenuto, eds. Basel: Birkhäuser.

————. 1997. *Epures D'architecture*. Basel: Birkhäuser.

————. 2009. Gaspard Monge Founder of 'Constructive Geometry'. Pp. 1293-1300 in *The Proceedings of the Third International Congress on Construction History*, Cottbus. http://www.formpig.com/pdf/formpig_gaspar monge founder of constructive geometry _sakarovitch.pdf (last accessed 30 June 2011).

STRUIK, Dirk J. 1933a. Outline of a History of Differential Geometry I. *Isis* **19**, 1: 92-120.

————. 1933b. Outline of a History of Differential Geometry II. *Isis* **20**, 1: 161-191.

TATON, Rene. 1951. *L'Œuvre scientifique de Monge*. Paris: Presses Universitaires de France.

————. 1966. La première note mathématique de Gaspard Monge. *Revue d'histoire des sciences et de leurs applications* **19**, 2: 143-149.

VELIMIROVIC, L. and M. CVETKOVIC. 2008. Developable surfaces and their applications. Pp. 394-403 in *Mongeometrija 2008*, Proceedings of the 24th International Scientific Conference on Descriptive Geometry, Serbia, September 25-27, 2008.

### About the author

Snežana Lawrence is a Senior Lecturer in Mathematics Education at the Bath Spa University, in Bath, England. She has been involved with a number of UK national initiatives to promote the use of the history of mathematics in mathematics education. Her website (www.mathsisgoodforyou.com) is a popular resource for secondary school children and teachers, attracting more than 60,000 page views a month. Snezana is interested in the historical issues relating to mathematics education in professional setting and as such has written and researched on the mathematical education deemed necessary for architects and engineers from the eighteenth century until today, in the countries of Western Europe and the United States. Snezana is on the editorial boards of *Mathematics Today* (a journal of the Institute of Mathematics and Its Applications, UK), and the *British Society for the History of Mathematics Bulletin*, and is the first Education Officer of the same society. Her chapter on the history of mathematics in the Balkans has appeared in the recent *Oxford Handbook of the History of Mathematics* (Jackie Stedall and Eleanor Robson, eds., 2008). As a member of the Advisory Board of the History and Pedagogy of Mathematics group (an affiliate of the International Commission on Mathematics Education) she is a regular contributor to their publications.

# Enrique Rabasa

ETS de Arquitectura de Madrid
Av. Juan de Herrera, 4
28040 Madrid SPAIN
enrique.rabasa@upm.es

Keywords: Jules de La Gournerie
Gaspard Monge, descriptive
geometry, stonecutting, vault
construction, skew arches,
perspective theory, axonometrics

Research

# La Gournerie versus Monge

Presented at Nexus 2010: Relationships Between Architecture
and Mathematics, Porto, 13-15 June 2010.

**Abstract.** Jules de la Gournerie was one of Gaspard Monge's
successors in the teaching of Descriptive geometry and its
applications, and both dealt with similar matters. La
Gournerie's writings criticized Monge's ideas with detailed
observations. In this article we go over all the issues on which
there was a marked divergence, from the stereotomy of stones
to the conical perspective or the organization of teaching. It
also show how the constant opposition between the two men,
even not explicitly focused on philosophical and political
questions, represents two radically different ways to
understand the world, each solid by itself.

## Introduction

At the middle of the nineteenth century, Jules de La Gournerie (1814-1883) was a
very important professor of descriptive geometry, a subject which had been founded by
Gaspard Monge (1746-1818) more than fifty years earlier, during the revolutionary
period. When Gino Loria mentions Jules de La Gournerie in his *Storia della geometria
descriptiva* [Loria 1921: 238-239], the only thing he remarks about him is his somehow
anti-Monge nature; after that, he moves on to explain his achievements for the science of
descriptive geometry. Jean Dhombres [1994] and Joël Sakarovitch [1998] have discussed
the general criticism that La Gournerie makes of the previous period. Dhombres focused
on the historical situation, the willingness of change of some engineers and the tendency
on the part of La Gournerie to opt for *une épistémologie du reel*.[1] Sakarovitch brilliantly
explains the role that La Gournerie plays in the history that binds stereotomy and
descriptive geometry. It is most likely that family circumstances and political ideas also
influenced La Gournerie's criticism towards Monge.

Here, I would like to provide some details about this criticism in some of the works
of La Gournerie, and show how radical his position was and how it became an obsession
during the rest of his life. But I would especially like to show that between Monge and La
Gournerie there are not only differences in opinions in the scientific field or regarding
the training of technicians and masons, stonecutting, or the basics of graphic systems of
representation, but that they also represent diametrically opposed conceptions of the
world, each of them consistent in itself.

## A universal graphic language

Gaspard Monge developed what he called *Géométrie descriptive* at the School of
Military Engineers of Mézières, where he taught for twenty years [Taton 1951]. Monge's
descriptive geometry focused on the use of horizontal (plans) and vertical (elevation)
coordinated projections to resolve spatial problems, up to that time known as the system
of double orthogonal projection, or simply method of projections. Since the Middle
Ages, the practice of stonecutting for construction purposes had grappled with complex
problems of this kind so as to shape ashlars and voussoirs that coincide. To achieve this,
the procedures used were similar to those of Monge, using a horizontal projection and

one or more vertical projections or *rabattements*. Monge systematized the procedures, reducing the tracing to two projections, stating the rules and establishing a vocabulary for the abstract geometrical elements, that is, points, lines and planes.

Due to his humble origins, Monge could not start his activity (which also comprised mathematics, physics and chemistry) in the more relevant scientific institutions. Even in the School of Mézières, he was forced to keep the graphic rationalization that descriptive geometry meant hidden, with the excuse that it had to be treated as a military secret. At the beginning of the French Revolution, in which he played a relevant role, Monge established the *École Normale* and the *École Polythechnique*. He explained his descriptive geometry publicly in both institutions.

Monge was a Jacobin who eventually became a Navy Minister. He took part in all the revolutionary initiatives related to technique and reason. He was part of the committee which instituted the system of weights and measures as well as its Republican calendar. He became involved in the rationalization of French industry, wrote about cannon manufacture, and so forth. He established descriptive geometry as a graphic, abstract and general language, and, as such, the basis for every kind of technical teaching.

The unification and legislation of graphic procedures by Monge were radical. In order to make it clear that the basic graphic and spatial thinking of descriptive geometry represented the language and foundation of all technical activities, the programme of the *École Polytechnique* presented not only stonecutting but also architecture itself as a part of descriptive geometry.

Most of continental Europe followed systems similar to those of the *École Polytechnique*. Descriptive geometry became an essential subject during the nineteenth century. Not only did the procedures of descriptive geometry resolve technical problems, but they also helped to develop rational geometry concepts; besides, the thought itself on the representation systems that use the idea of plane projection was of great interest from the viewpoint of geometry. As a result, descriptive geometry as a part of geometric science advanced strongly during the nineteenth century. Thus, although it was soon evident that the privileged position granted to this subject by Monge within the general education of engineers and architects was somehow out of proportion, it retained an important position in the programmes of technical schools up until the end of the nineteenth century. In fact, in order to make up for the growing importance of mathematic analysis, descriptive geometry became more and more abstract, especially during the second half of the nineteenth century. It was based on Jean-Victor Poncelet's theory of projective geometry; Poncelet, a follower of Monge, explained the general properties of the plane projection at a conceptual level.

Even though descriptive geometry was first explained publicly in the *Écoles Normale* (it was taken down by stenographers for prompt publication in 1795), its natural place was the *École Polytechnique*. After Monge's death, his followers at l'X (the abbreviation used to indicate the École Polytechnique), especially Jean Nicolas Pierre Hachette and Charles François Antoine Leroy, continued teaching the subject, but the mathematic analysis were increasingly emphasized (by Pierre-Simon Laplace and Augustin-Louis Cauchy). In the *École Centrale des Arts et Manufactures* and the *Conservatoire des art et Métiers*, another of Monge's followers, Théodore Olivier (1793-1853), wanted to keep Monge's spirit alive by privileging graphic thinking. Olivier wrote several interesting treatises in which he always highlighted the concept of descriptive geometry as a common language for all kinds of engineers, the vehicle of their spatial and constructive thinking

However, Olivier himself began introducing certain notions on a different system of representation, axonometry [Olivier 1852: 123-129].

Jules de La Gournerie replaced Leroy in the *École Polytechnique* in 1849, and Olivier in the *Conservatoire* in 1854. To this aim he presented a *Discours sur l'Art du Trait et la Géométrie descriptive*, which is a short history of the systems of the space representation [La Gournerie 1855]. In 1860 he published his *Traité de Perspective linéaire*; in 1865 his *Géométrie descriptive*; and in 1874 a "Mémoire sur l'enseignement des arts graphiques", which is of great interest to us. He didn't get to publish a planned *Traité de stéréotomie*, but he wrote about a problem that was important at that time, the stereotomy of oblique arches. It is true that his total rejection of everything Monge did is much more obvious in his last work, although when analysed carefully, there is no doubt that it is present since the beginning of his work; in fact, his *Discours...* of 1855 is understood in a different manner after reading the *Memoire...* of 1874. Indeed, the selected themes in this *Discours...* make more sense in the light of the more freely expressed criticism exposed two decades later.[2] For this reason, and because we are interested in emphasising the differences between these two men rather than their biographies, we will review the work of Jules de La Gournerie in a synchronic manner.

However, it should be taken into account, regarding the lives of these two important figures, that Monge was a scientist, very interested in the practical application of scientific findings, as well as a professor of engineering who never actually built anything in his life, whereas La Gournerie was a brilliant and efficient engineer, forced *par devoir* to teach [Dhombres 1994], and the author – probably out of a sense of duty to demonstrate his academic ability – of several monographs about specific geometric themes that are impeccable from a scientific point of view.[3] It could therefore be said that each lived a life which, in some way, did not correspond to him.

## Stonecutting and Monge's system

Monge himself admitted that the origin of descriptive geometry lay in the common practices of the construction trades and, especially, the tracing of stonecutting. However La Gournerie says that this new discipline had been presented as if before Monge there was only disorder and confusion [La Gournerie 1874:153]. The *Discours...* (a title which already justifies the range and antiquity of the procedures of technical drawings of various trades and activities against the self-limitation of descriptive geometry to one system only) shows a direct knowledge of the old ways of stonecutting. La Gournerie [1855: 9] quotes the following paragraph written by Philibert De l'Orme (1567):

> *J'emploirai le temps que me será plus à propos à revoir Euclide et acommoder sa théorie avec la practique de notre architecture, lui accompagnant Vitruve et le réduisant à une certaine méthode,*

and explains that De l'Orme died before he could finish his project. He would quote exactly the same paragraph twenty years later, in his "Mémoire sur l'enseignement" [La Gournerie 1874: 124], to specify that the first question, the geometrical explanation of graphical procedures – to quote Euclid here is to invoke the geometric science – had been resolved by Frézier's treatise on stereotomy [1737-1739], while the second one, the writing of a reasonable and definitive stereotomy, had not been attained yet, and Monge had moved away from this purpose, because *il a abandonné Vitruve*, that is, he has forgotten the art of construction.

In the surveys of the stereotomy treatises previous to Monge, La Gournerie carefully emphasises that the arrangements used in construction had changed very little, except those that present warped surfaces [La Gournerie 1855: 27]. Although he does not mention it, it was well known that the most remarkable advances in the knowledge of warped surfaces are due to geometries posterior to Monge, such as Hachette and Chasles.

La Gournerie knew the work of the mathematician Girard Desargues well. Desargues had considered various arrangements in which the cylindrical surface of a barrel vault terminates by being cut by one plane, which is the face of a wall. But he did it in an abstract way, moving the geometric elements (cylinders and planes) to a position in such a way that operations were more general. This abstraction is natural for us, but for those who had to cut stones, the natural thing was to understand that every case was different, depending on whether the vault was horizontal, sloping or oblique; or whether the face of the arch was vertical or sloping, and so forth.

Desargues's proposal was radically rejected by those who dedicated themselves to stonecutting, and La Gournerie understood why. The *appareilleur*, the tracers or master masons, work with representations of material elements that they must transmit appropriately to their subordinates and, especially, to those who cut stone. During this process, it is convenient that the horizontal planes remain horizontal, and the references of the graphical traces used maintain an immediate relation with reality. This has always been done, and while keeping up with this tradition is irrelevant in the field of geometry, it is not so for a technician.

In the teachings of the system of double orthogonal projection that he developed in the *Conservatoire*, the pro-Monge Olivier had inserted the change of projection plane; that is, like Desargues, he wanted to change the position of the geometric elements in geometrical problems. La Gournerie was against the changes in projection suggested by Olivier [La Gournerie 1855: 44-46; 1860: VI-VIII; 1874: 154], and referred to the experiences of the past.

It is in relation to this matter that he expressed his opinion about Monge's School with more cruelty. La Gournerie wanted to show that Olivier and Leroy were not aware of, or had not carefully read, the earlier authors. La Gournerie explained that Leroy had praised Frézier and emphasised that he had already used two projections, following Monge's style, without noticing that La Rue (1728) had done it before. Then, La Gournerie adds sardonically that if Olivier *avait suivi le sage conseil de Leroy* [La Gournerie 1874: 154], if he had at least read Frézier, he would have found the remarks made concerning Desargues's procedures, and he would have known that this issue about the changes in the plane of projection had already been discussed in relation to the stereotomy field, and had been rejected.

This procedure – the change of the plane of projection, consisting of obtaining the shapes and surfaces by means of an abstraction, making them independent from the real arrangement of the pieces where they are to be found – could only be followed in the case when the workers had been trained in the procedures of descriptive geometry with the same knowledge as the engineers and architects. This represents another disagreement of La Gournerie with the Monge School, which we will discuss later.

In general, it was easy for La Gournerie to demonstrate that the previous traditions related to what was to be considered the origin of descriptive geometry – that is stonecutting – were very rich and diverse. After this, the figure of the famous Monge would be definitively discredited if it were possible to demonstrate that his concepts in relation to the practical applications are misleading, and that the organisation that he imposed on the teaching had led to the decline of the French stereotomy. La Gournerie would enthusiastically devote himself to this issue.

## Lines of curvature

Monge discovered that algebraic surfaces implied an interesting property. On any surface of this kind, there are two families of curves that cross each other orthogonally and that we can trace if we take the direction of the maximum and minimum curvature on each point. Let's imagine the straight line normal to the surface in every point of the curvature lines. These normal straight lines form ruled surfaces. Then, there are two families of ruled surfaces, orthogonal to each other and orthogonal to the first surface. Monge discovered that these ruled surfaces are also developable. Let's see how such an abstract proposition can be related to the practice of stonecutting.

There are two important rules in stereotomy that, up to that time, were not usually explicitly stated, although these are obvious to those who know about the art of construction and stonecutting. The first rule is that acute angles should be avoided. An acute angle is a delicate point in the process of cutting and during transport, and is also a weak point for receiving forces. In a segment of a vault, the meeting point between the lateral joint with the surface of the intrados should not be acute. However, if it is obtuse, the acute angle will appear in the adjacent piece, so it is generally desirable that this meeting point should be orthogonal.

The second unstated rule is that the surfaces cut by the stonecutter should be easy to check. To cut is to remove material and, during this process, it is essential to check continuously to make sure that the result is correct. The stonecutter can easily check the execution of a plane, of a cone or of a cylinder, as they are surfaces generated by straight lines. In addition, these surfaces also have the advantage that it is possible to make a template, rigid for the plane and flexible for the cones and cylinders, that can be laid out on a surface in order to check the perimeter of the cutting face. So that the advantage of using developable ruled surfaces is evident: a flexible template can be laid out on them.

Monge found that his theory of curvature lines fully satisfied both conditions, which are exclusively of a geometrical or formal nature [Monge 1799: 106-126]. Naturally, the design of an arch or vault should correspond to other, specifically mechanical conditions. But knowledge of the statics of structures was based more on experience than on calculation, and the treatises of stereotomy did not usually deal with geometrical and statical problems in the same place.

Monge's concern to find practical application to the scientific findings led him to propose the quartering according to the curvature lines of all type of vaults (fig. 1). However, he especially insisted on the suitability of designing the elliptic-plan vaults and ellipsoidal surfaces this way, and he even recommended that this should be the shape of the hall for the debates of the *Assemblée Nationale*, for which an architectonical competition was announced during the third year of the Republic.

The theory of curvature lines and the Monge's opinion about the quartering in stereotomy is reflected in several of his writings;[4] the proposal of replacing the traditional quartering of elliptic vaults – done by horizontal courses – with the quartering according to the curvature lines, was followed by his disciples. Monge's quartering for the elliptic vault appears frequently in the books of stereotomy of the nineteenth century (figs. 2, 3). The reader of those treatises may conclude that this construction of the vault is a real alternative, but the truth is that it is so complex (in the sense that the resulting pieces, complying with those ideal conditions referred to orthogonality and developable ruled surfaces are so sophisticated) that as far as we know it has never been constructed. However, La Gournerie does not mention this fact, so he may have thought that it might be still possible to carry it out.[5]

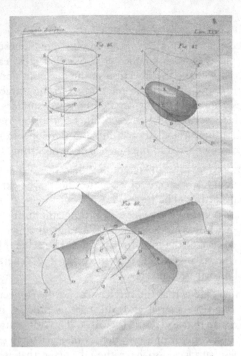

Fig. 1. The lines of curvature in Monge's *Géométrie descriptive*, 1799

It was extremely easy for La Gournerie to emphasize Monge's enormous mistake by enunciating, in a general way, a law that can in fact only be complied with in a very few cases, and follows only geometrical considerations. Monge's generalization does not take material circumstances into account, nor does it warn about the fact that there are cases in which a certain tolerance with acute angles is inevitable and others, in which the demand that surfaces be mathematically developable is excessive; it does not distinguish between beds (bearing pressures) and joints (that simply separate); it presupposes an absolute perfection in the cutting and neglects the presence of the mortar between the pieces; it does not take account how the vault is terminated, or its intersections with other vaults. Moreover, and from the geometrical point of view, it is not strictly true that the orthogonality between ruled joints and the intrados demands the use of adjusted lines to the intrados [La Gournerie 1874: 128-129].[6]

La Gournerie mentions several times in the *Discours...* that stereotomy had improved little since the sixteenth century and cites as exceptions those cases in which the theory of the ruled surfaces should be present. A characteristic example of this is the famous *Arrière-voussoir de Marseille*, a way of covering a door by a splay whose intrados is a ruled surface. There is a problem, because the conditions that define the surface are not equal in all zones, and this means that, in fact, there are two different adjusted surfaces, one after the other. Without the proper precautions in the design, it may turn out that instead of a perfect continuity and tangency between the two zones, there is a groin. However, La Gournerie says, if this should occur, the problem would be easily solved in the stonecutting process itself; he is quite right in this and in fact the previous stonecutting treatises do not even warn that this could be a problem [Sakarovitch 1998; Rabasa 2000]. As has sometimes been pointed out, a problem is not considered as such until there are historical conditions for its solution.

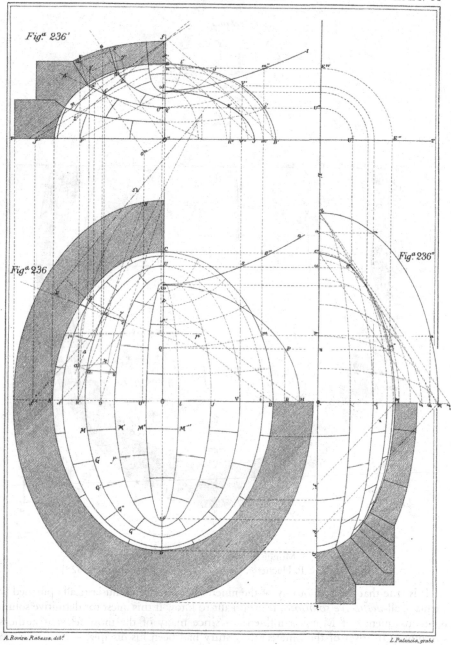

Fig. 2. Monge's arrangement of the elliptic vault
from Antonio Rovira, *Estereotomía de la piedra* [1899]

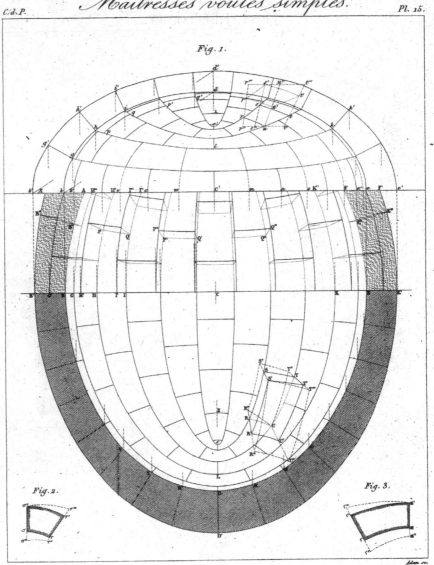

Fig 3. Monge's arrangement of the elliptic vault
from Jean N. P. Hachette, *Traité de Géométrie descriptive* [1828]

It is true that the stereotomy of the nineteenth century enthusiastically pursued wha
we may call *definitive solutions*. It is difficult to know if this quest for definitive solution
is a consequence of Monge's influence – since many of the most relevant authors o
stereotomy treatises of the nineteenth century had been his disciples – or if it is simpl
the result of the faith placed in the scientific progress in a century in which the geometr
of three-dimensional space was brilliantly developed to its logical conclusion. Althoug
iron in constructions was slowly introduced during the nineteenth century and concret
at the end of the century, structures were frequently made of stonework, and the ston

stereotomy was still useful, so it is not strange that a brilliant application of geometry to stereotomy was expected. This application turned out to be a bit unnatural in cases such as the quartering of the elliptic vault according to the curvature lines, and the *Arrière-voussoir of Marseille* – or Abeille's flat vault in the eighteenth century –, in which the previous uses, which had been sufficient in practice, were proposed to be changed by others that would definitively resolve the supposed problems. As we have seen, the the presence of these problems had to be emphasized, in spite of the fact that they were irrelevant from the standpoint of construction.

La Gournerie notices some of this when he mentions, as commented above, that the only real advances of stereotomy in the nineteenth century are those related to ruled surfaces. This means that the work of architects and master masons during the previous centuries had developed a discipline that needed no improvement. However, there is another type of arrangement that represents an exception in this scene, insofar as it was not used with the same frequency: the large oblique arches for railway bridges. The railway lines could not adapt their routes to meet paths and rivers in an orthogonal manner and consequently the required bridges, when made of stonework, presented large skew arches. The design of skew arches in relation to the faces of walls that they went through had been previously resolved for the small scale of architecture, but these solutions were not sufficient in the case of large bridges. The stereotomy of the nineteenth century dealt with this problem, arriving at two solutions of great interest [Rabasa 1996; Sakarovitch 1998] (figs. 4, 5).

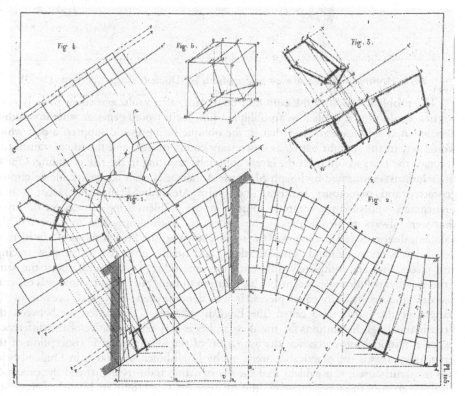

Fig. 4. *Appareil orthogonal* to skew bridges from J-P-Douliot, *Coupe des pierres* [1869]

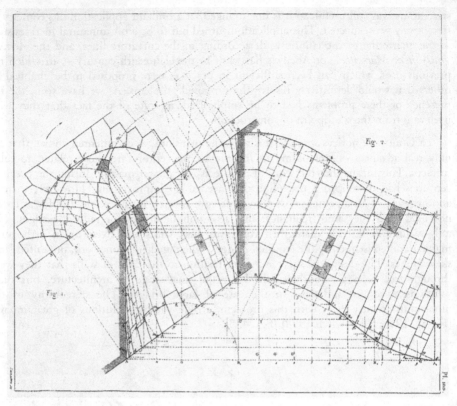

Fig. 5. *Appareil helicoidal* to skew bridges, from J-P-Douliot, *Coupe des pierres* [1869]

The problem was that the common quartering of a vault, according to horizontal courses, could not be applied to an oblique vault, for it would generate what it was then denominated as *poussé au vide*; that is, the oblique facing would suppress a part which would act, in the straight barrel, as a necessary counterthrust, and the forces transmitted amongst the rows would meet the layers obliquely, resulting in the risk of sliding. One of the solutions, enunciated by Joseph Alphonse Adhémar (author of remarkable descriptive geometry and stereotomy treatises) and Pierre Alexandre Francisque Lefort – both Frenchmen – pursued the complete solution to the problem using surfaces for the layers that were always perpendicular to the forces. Is almost impossible to carry out this solution, denominated 'orthogonal' (it is also known as 'the French solution') in its most detailed version; it demands simplifications, and finally, gives rise to complex tracings and pieces that are completely different from one other. In England, about the same time, another solution was reached. This solution did not pursue perfection, was easier to execute, and resulted in equal pieces; it was denominated 'helicoidal' because of the pattern of the joints (also called 'the English solution'). The difference between the French and English solutions for the skew arches is very representative of the difference between national idiosyncrasies: the aspiration of the ideal geometric conception of the Frenchmen versus the practical concern of the Englishmen. Although in England brick was frequently used, it is evident that the construction industry demanded the execution of arrangements in stone; however, the treatises on stereotomy in English, before and after Monge, hardly existed; there is no need, in fact, to have a complete and precise theory to resolve occasional problems of this sort.

Fig. 6. Helicoidal arrangement in the Puente de los franceses, Madrid

The personality of Jules de La Gournerie was clearly closer to the English way of thinking than the French one. Practical solution, instrumental character, the concern with 'making' as opposed to 'naming accurately', the respect for the procedure reached and ensured by tradition, brings La Gournerie near to the Anglo-Saxon methodology. He presents the advances coming from several places abroad as a consequence of the clear decay of the French style that Monge started.

Regarding oblique arches (and to defend himself in the controversy that he had with Lévy, who was an engineer as well), La Gournerie wrote a *Mémoire sur l'appareil de l'arche biaise* [La Gournerie 1872], in which, in addition to presenting his opinion in relation to this matter, he took advantage of the opportunity to reiterate the criticism of Monge's application of the curvature lines to stereotomy. (The controversy between La Gournerie and Lévy is well detailed and elegantly studied by Antonio Becchi [2002].)

## Methods of teaching

La Gournerie's "Mémoire sur l'enseignement des arts graphiques" [1874] although it repeats many subjects already detailed in the 1855 *Discours...* and other writings, gives an account that shows how, despite Monge's imposition, some engineers maintained a certain common sense. We are referring to the attitude of Gaspard Riche de Prony (1755-1839), a prestigious engineer who was the director of the *École des Ponts et Chaussées* during the same period in which Monge and his disciples obliged students to first attend the *École Polytechnique*.

La Gournerie based his research on a careful reading of the manuscripts that Prony left, and he reached the conclusion that Prony defended the practice of teaching stone stereotomy in connection to the construction practice, whereas Monge's School insisted in maintaining at the École Polytechnique the earlier custom of teaching a geometric and abstract stereotomy. This decision was consistent with the exclusively geometric conception of stereotomy that emerged in Monge's writings, and made it possible to present the system of double orthogonal projection as a universal language and illustrate its ability to solve the most complex spatial problems.

La Gournerie's style was limited because of his concern for meticulously analyzing and carefully demonstrating all his assertions. Despite this, in the "Mémoire sur l'enseignement" [La Gournerie 1874: 115], he allows himself a literary resource: he begins narrating Prony's difficulty before Monge and his disciples and, after a detailed review of Monge's erroneous criteria and their consequences, he resumes this subject at the end of the book, explaining that, despite the passage of time, the elderly Prony never changed his mind.

What Monge was seeking was a unified concept of spatial thought, one that was the same for all the various kinds of design activities. He also sought to satisfy the encyclopedic demands of the period by providing explanations of the methods of various crafts and industries. Scientific progress can only benefit all forms of crafts and industry if the basic procedures are public and common. Monge thought that such a centralised organization, which would select the best results from an earlier period of preparation, would only be possible if the closed, secretive atmosphere characteristic of the trades was abolished. La Gournerie, in contrast, believed that the specific traditions and abilities, the procedural differences between craftsmen and designers which are the consequence of refined experience in the work, would be lost with such a *tabula rasa*, and that only loss of richness and diversity could derive from an abstract unification.[7]

According to La Gournerie, the engineers in the *École Polytechnique* learned an ideal stereotomy that they could not criticise because they didn't have the necessary constructive knowledge [1874: 123]. The craftsmen, the masons, were not able to achieve the quality obtained in the past, because the learning system characteristic of the *compagnonnage*, that is, the apprenticeship under the master mason, had been abolished.

La Gournerie says that trades such as carpentry (which had evolved without being influenced by Monge's ideas) and perspective drawing (which was a thing for artists) had been affected to a lesser extent.

### Linear perspective

Desargues also proposed a *trait universelle* for perspective; as in the case of stereotomy, Desargues proposed a method that was perfectly alien to the evolution of the procedures of the perspective drawing, starting from scratch with an abstract idea. La Gournerie made it quite clear in his *Discours...* that Desargues's procedure for perspective had also been rejected or ignored by later authors. In an ironic fashion, which the reader in that period probably did not notice, he includes Monge amongst those who rejected Desargues's abstract and mathematical concepts [La Gournerie 1855: 20, note] It is no surprise that Desargues – like Monge, a scientist who tells the technicians what to do, although further removed than Monge from the practice he wants to modify – suffered the radical rejection of Jules de La Gournerie.

Monge was not especially interested in perspective, although, as this was a subject to be included in the training of engineers and architects, and he himself had proposed the system of double orthogonal projection as a basis of all particular graphic methods as universal language, he necessarily had to give some account of it.

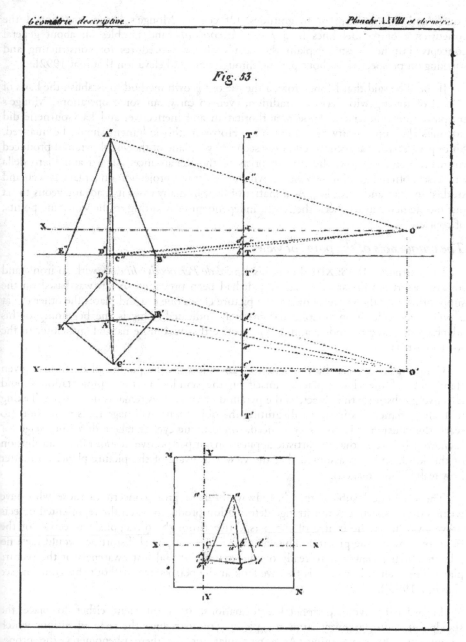

Fig. 7. Plate about perspective, from Monge's *Géométrie descriptive* [1820]

Monge proposed tracing conical perspectives by finding the position of the elements by means of the system of double orthogonal projection, using plan and elevation to locate the point of view and the picture plane, and to project the elements of the object on it; in other words, he proposed finding a conical projection by means of orthogonal projection. Let's remember that in that period the theory of perspective had found procedures and resources internal to perspective itself, developing the use of vanishing points and the measure of magnitudes. Of course, Monge's lessons mentioned the concurrence of parallel lines in a *point de concours*, and rambles on about general concepts, but he doesn't explain the details of the procedures for constructing and working on perspective without passing through plan and elevation [Laurent 1992].

It could be said that Monge forced the use of his own method to establish the basis of a kind of tracing with its own tradition, even to carry out some operations. Monge's purpose, thus enunciated, appears authoritarian and ineffective, and La Gournerie did not miss the opportunity to present it so. However, this judgment should be nuanced. Since perspective was recognized as the section by a plane of the visual pyramid produced from the point of view, the first theorists of the Renaissance, Alberti and Piero della Francesca, offered graphic sketches in which this central projection on a plane is seen and studied in plan and elevation. As a matter of fact, in many cases it is advantageous to set out the elements as to check the result, independent of resorting to the vanishing points, of course.

## The uniqueness of the point of view

La Gournerie [1858: XIII] began his *Traité de Perspective linéaire* with an ironic and forceful assertion: linear perspective, as it had been presented so far, was based on the supposition that the viewer looking at a picture closes one eye and places the other eye at a point whose location no one has precisely indicated. This is the beginning of his advocacy of a way of understanding perspective that was more natural and closer to the work of artists.

Conic perspective is the projection of an object on the picture plane from a given standpoint. If the viewer places himself on the standpoint, the representation should effectively substitute the object, as the pyramid of visual rays remains unchanged. Taking it to an extreme, this idea of substituting the object with its image leads to illusion (to reach this illusion it is necessary, indeed, to close one eye, or take a different image for each eye). It is clear that the artistic applications of perspective do not pursue an illusion of this kind, and it is assumed that the viewer is aware of the picture plane, no matter how realistic the image is.

The theoretical subject that has always been of great concern to those who have written on perspective is the strange deformation produced when the represented objects move away from the centre of the perspective (from the orthogonal projection of the point of view on the picture). This deformation, a peripheral distortion, would have no relevance if the viewer were really on the standpoint and lost awareness of the picture plane. The reality, however, is that we look at the perspectives without this concern (see [Pirenne 1970]).

In order to avoid peripheral deformations, it is sufficient either to place the standpoint quite far away from the picture, or to reduce the field of vision, which amounts to the same thing. As a particular case of these phenomena, the proper perspective of a sphere a bit away from the centre should be an ellipse; moreover, if the perspective of a frontal colonnade is properly traced, the columns should appear thicke as they move away. Evidently, the artists had never been so rigorous, and in hi 1858 *Traité de perspective linéaire* (but not in his 1874 "Mémoire sur l'enseignement" La Gournerie observed that, in practice, several points of view can be admitted, i necessary; that is, that the theoretical principle of the uniqueness of the point of view can be disregarded.

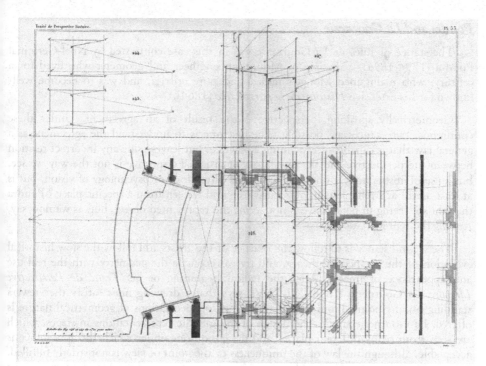

Fig. 8. Plate 35 from Jules de La Gournerie, *Traité de perspective linéaire* [1858]

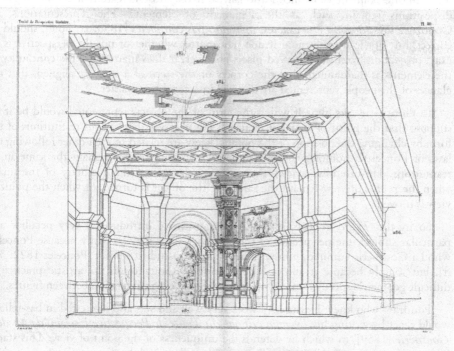

Fig. 9. Plate 40 from Jules de La Gournerie, *Traité de perspective linéaire* [1858]

## Poudra and La Gournerie

The stance of Jules de La Gournerie was in this case contrasted by Nöel-Germinal Poudra (1794-1894), a disciple of Monge, an engineer and geometer who lived for a century, who maintained Monge's radical and pure criteria, and was to become well-known for his extensive *Histoire de la perspective* [1864].

Geometrically speaking, perspective is the result of an agreement. Under these conditions, two stances are possible. The first one is to understand the agreement as a general law that is capable of explaining a practice, but leaves aside any incorrect relation between the perspective and the physiology of vision. Perspective is not the way we see, but a useful abstraction; it does not reproduce physiology or psychology of vision, but is at least useful as a reliable map of what is seen and unseen from a specific place.[8] Poudra thought so: seeing in perspective is not *seeing* the represented object, but, as we now say, *reading* the representation.

The second stance is to follow the practice of the artists and follow the slow historical evolution of the perspective theory, and try to reconcile the geometry with the real use and appearance of images. After the radical beginning of his *Traité de Perspective Linéaire*, La Gournerie [1858: XII-XXVI] says that a drawing must satisfy the viewers standing on any point. From here, a wide range of speculations of geometrical nature is offered. La Gournerie studies the design of the graphic representations in stages, which are seen from many different places, and the conditions under which the appearance is acceptable, although the law of the uniqueness of the point of view is not strictly fulfilled. In order to arrive at an acceptable appearance there is no need to obey the requirement for a unique point of view unconditionally, but only to take care that certain apparent dispositions remain, such as the alignment of elements. The representations, La Gournerie says, *doivent être exactes, mais non pas complètes* [1858: xv, note], should be correct, but not lead to illusion: it does not matter whether or not the perspective is an exact projection from the observed place; however, it does matter that the continuity of the elements is maintained, that the corners of the steps of a stair are aligned, that the glances of the people represented are directed in a consistent matter.

La Gournerie asks himself which the form of the objects represented would be if we suppose that the point of view has changed. He observes that various restitutions of the form by changing the position of the point of view are related to each other following the laws of Poncelet's homology. The continuities and alignments remain the same in all restitutions. He carefully reviews the apparent changes of the geometry of the image when the points of view remain the same but the object is turned, or when the point of view is moved.

So in the *Traité de Perspective* de La Gournerie introduces a very peculiar and particular subject, the perspective of the bas-relief. This may be partly because Poncelet who La Gournerie admired, hinted that it was an unstudied subject [Poncelet 1822: 36? ff], but mainly because it allows him to connect thoughts on the artistic practice to difficult geometric speculations by means of the homological relation between figures.

Poudra – who knew Desargues's work very well and was also interested in bas-relief – would react by writing his *Examen critique du Traité de Perspective Linéaire de M. de* Gournerie [1859], in which he defends the uniqueness of the point of view. This stance is not as dogmatic as it may seem. To admit that the principle of projection from a point is a conventional law makes it possible to encompass a wide range of cases. In fact photography has allowed us to see rigorous perspectives that would have been a bit strange for the traditional view of La Gournerie. It can even be maintained that it is more honest to accept that this is only an agreement – and observe the adaptation of the view to the socially established language – rather than continue to seek a confused equilibrium between geometry and perception, a natural justification. Poudra is not very brilliant in his defense of the School of Monge, although he is right about the conventionality of graphic representation.[9]

## Axonometry

If La Gournerie claimed the plurality of the graphic procedures to represent and work with three-dimensional space, he could not avoid making references to axonometry.

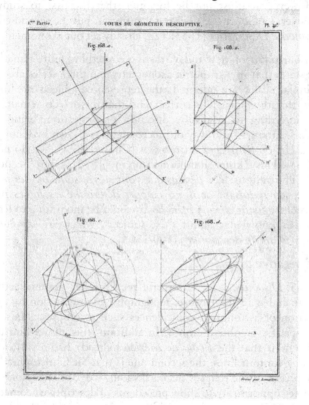

Fig. 10. Plate about axonometry. Théodore Olivier, *Cours de Géométrie descriptive*, 1852

Monge's School did not take axonometry into account. It should be borne in mind that the use of axonometries that are orthogonal projections (that is, not those that are oblique projections), began with the work of the English author William Farish [1822]; during Monge's period only *cavalière* or *militaire* perspectives were used. Obviously these are oblique projections, although they were not often seen as such. We frequently find comments that show that these models of perspective were thought of as conic perspectives in which the convergence of the lines and the diminution of the sizes are not rigorously observed.[10] La Gournerie even refers to all axonometries as forms of *perspective rapide*; he in fact uses the word axonometry only for the orthogonal perspective, as if cavalier perspective was not based on axes.

However, Monge went so far as to delete cavalier perspectives from the stereotomy illustrations taken from De la Rue's treatise on stonecutting [La Rue 1728] for the exercises of the *École Polytechnique*, so that no manner of representation could hinder the efficacy of the system of double orthogonal projection.

Concerning this subject, La Gournerie complains once more about how French studies lagged behind those being carried out in England and Germany, which were developing orthogonal axonometry, and he introduces certain rudiments in his *Traité de Géometrie Descriptive*.

It has sometimes been said that Auguste Choisy, who introduced axonometric representation in the analysis of historical architecture, had studied with Jules de La Gournerie. However, it must not be forgotten that the first to study this subject in France was Olivier. La Gournerie does not mention it, but the second edition of Olivier's *Cours de Géomètrie Descriptive* [1853: 122] already set out axonometry (fig. 10).

Seen from our point of view today, there is something quite remarkable in Olivier's work. Olivier states that in a trimetric axonometry the three scales are different and that if, instead, similar scales are adopted, the represented objects are different. Pohlke's theorem shows that this practice is correct, and that the objects remain exactly so if they are projected according to a specific direction of projection, being oblique and not orthogonal. This extremely odd way of proceeding was chosen by Choisy for some of the figures of his *L'Art de bâtir chez les romains* [Choisy 1873]. We do not know if at the time of writing Choisy knew Pohlke's theorem (1858),[11] but he nevertheless writes concerning the illustrations: *Les dessins que vont suivre sont, ou des projections, ou des figures obtenues en réduisant dans un rapport déterminé les lignes parallèles aux axes gradués qui accompagnent chaque planche.* It would be interesting to learn the real scope of La Gournerie's knowledge about these subjects, which are set out in a very basic manner in his *Géométrie descriptive* [1860-1864].

### Positive and negative utopia

In the 1855 *Discours...* La Gournerie recalls other graphic activities that could reasonably be included in the treatise: topographic representation by contour lines, and graphic deduction of astronomic coincidences such as eclipses. Thus, he broadens the variety of the specific graphic methods. In addition, this allows him to emphasise two facts: the first one is that the *École de Mézières* already had a sufficient knowledge of representation by contour lines; the second one is that Sir Christopher Wren was able to determine the phases of the eclipses thanks his knowledge of stereotomy. It could be said that he took every opportunity to show precedents of descriptive geometry. When quoted by La Gournerie [1855: 24, note], a specific sense is acquired by the phrase that Lagrange said after hearing a lecture given by Monge: *Je ne savais pas que je savais la Géometrie descriptive* ('I had not realised that I knew descriptive geometry'): the meaning is not that descriptive geometry is a natural language, but that Monge's originality should be questioned.

La Gournerie stood almost alone in his radical opposition to Monge, though others shared, to a greater or lesser extent, some of his criticism. Examples are Louis-Lège Vallée, a quite intelligent and shrewd engineer who could have held the professorship of descriptive geometry in the *École Polytechnique*, and Poncelet, who was his director when Leroy is replaced by La Gournerie. Both were Monge's disciples and while they probably noticed his weak points, they also recognised the general boost that he gave to the rationalization of the graphic methods. Prony should also be included, of whom La Gournerie is obliged to explain the passivity he showed at the end of his life. La Gournerie himself tempers his scathing irony with a certain prudency, both at the start of his academic career, in the 1855 *Discours...* where he praises his predecessors Leroy and Olivier, and at the end of the "Mémoire sur l'enseignement" [1874: 156] when he only invites a certain reflection, shortly before moving into active politics as a moderate Catholic.

La Gournerie was the son of a monarchist combatant, and he inherited the title of viscount. Monge was first a revolutionary and then a friend of Napoleon, who granted him the title of Comte de Pélouse. We could probably find many personal reasons for La Gournerie's radical opposition to Monge, but the most interesting thing about this antagonism is the manner in which each of these two poles represent a consistent stance in relation to the technical creation. The stance is typically French in the case of Monge, in its pursuit of the ideal solution, and a bit more Anglo Saxon in the case of La Gournerie, who accepts tradition. More generally, their confrontation is expressive of the contradiction between reason, geometry, law and unity on the one hand, and history, technique, agreement, diversity on the other; or, seen another way, between utopia and nostalgia.

## Acknowledgments

This article is part of the research project "Ashlar construction in the Mediterranean and Atlantic areas. Analysis of built examples", BIA 2009-14350-C02-01, which is funded by the Ministry of Science and Innovation of Spain.

## Notes

1. Dhombres [1994: 18] stated that the criticism of Jules de La Gournerie is not directed personally to Monge, but to a general situation. It may seem odd to call La Gournerie's opposition as "personal", since Monge died when La Gournerie was only four years old. But the truth is that his criticism is totally directed towards a person, what he did, what he said and what he represented and, in an indirect manner, towards those who adopted a similar stance.
2. The praise that he is forced to make in 1855, when the figure of Monge was irrefutable, focuses on aspects of rationalization in relation to previous procedures as superfluous as Monge's adoption of a proper vocabulary: *projections, traces, lígne du terre*. However, it is not true that Monge used the expression *lígne du terre*, 'ground line' (see [Loria 1921: 107]); Louis-Leger Vallée [1819] was the first one to transfer it from perspective to the system of double orthogonal projection, and Adhemar [1841] expressed his opposition, precisely by invoking Monge's authority.
3. Especially *Recherches sur les surfaces réglées tétraédrales symétriques* [La Gournerie 1867], and briefer studies about the rotation of lines and second-degree surfaces about an axis.
4. Appears the first time in [Monge 1796], and after that in [Monge 1801: 17-21] and its reissues as *Application de l'Analyse à la Géométrie*, but also in the writings of Hachette, Vallée and Leroy.
5. Another arrangement based on abstract speculation and alien to practice is Abeille's flat vault. La Gournerie refers to him in a contemptuous manner, and says that this kind of vault has never been constructed. However, there were at least two flat vaults in Spain at the time he wrote this [Rabasa 1998].
6. La Gournerie noted that when tracing any curve on the intrados, the enveloping surface of a plane tangent to this line and normal to the voussoir in its meeting point is a ruled surface that meets the intrados at a right angle.
7. It is true that the technical literature suffered an important interruption. Before Monge, the authors of stereotomy books were aware of the previous treatises. After Monge, they behave like scientists and are only interested in mastering their immediate predecessors. It is striking that a

widely used book published at the end of the nineteenth century, the manual for exercises in descriptive geometry by Frère Gabriel-Marie, states that the old and almost legendary *biais passé* is a surface invented by Hachette (1769-1834), a disciple of Monge [F. J. 1893: 425, note].

8. It could be said that this is how August Choisy uses perspective. His relationship with La Gournerie will be commented below. In his *Histoire de l'Architecture* [1899], Choisy presents perspectives such as the view of the Acropolis from the Propylaea, in which the peripheral deformations are not corrected, and the result is more a map – as E.H. Gombrich would say – of what one can see from there, rather than the registering of a perception.

9. Another interesting debate, although a bit off this subject, is the one between Chasles and Olivier, when the former states that a tracing made with descriptive geometry is not capable of establishing if a curve is plane or warped. La Gournerie enters in the debate and concludes that it is a question of names, as it can be supposed that the Descriptive geometry comprises the general rational geometry applied to the tracing, and this allows knowing that circumstance. Although, probable, what Chasles wants to say is, as for language, the Descriptive geometry represents only forms and specific spatial situations, whereas the common and algebraic language can refer to general or familiar concepts. In relation to this, one should stress how La Gournerie seem to have no interest in the idea of the Descriptive geometry as a logical language, *la langue de l'homme de génie*, in the expression of Monge that so has impressed to Adhemar, Olivier, Leroy and others, who repeat it, adding puns between *écrire* and *décrire*.

10. For instance, see the contents of the manuscript from Mézières in [Olivier 1847: 9].

11. Pohlke enunciated his theorem (three segments concurrent of a plane could be always considered as a projection of the three concurrent edges of a cube) in 1858, although it was Hermann Amandus Schwarz who proved it in 1863; see [Loria 1921: chap. XII].

### References

ADHEMAR, Joseph. 1834. *Traité de Géométrie descriptive*. Paris: Fain et Thunot.

BECCHI, Antonio. 2002. *Chambre H*: Per una storia del construire. Pp. 17-100 in Antonio Becchi, Federico Foce, *Degli arch e della volte: Arte del construire tra mecanica e stereotomia*. Venice: Marsilio Editori.

CHOISY, August. 1873. *L'Art de bâtir chez les romains*. Paris: Ducher et Cie.

——. 1899. *Histoire de l'architecture*. Paris: Gauthier-Villars.

DHOMBRES, Jean, (ed.). 1992. *L'École Normale de l'an III, Leçons de Mathematiques, Lagrange-laplace-Monge*. Paris: Dunod.

——. 1994. La Gournerie, Jules Maillard de (1814-1883), professeur de géométrie descriptive (1845-1883). Pp. 18-35 in *Les professeurs du Conservatoire National des arts et métiers dictionnaire biographique, 1794-1995*. C. Fontanon, A. Grelon, eds. Paris: INPR-CNAM.

DOULIOT, Jean-Paul. 1869. *Coupe des pierres*. Paris: Dunod.

F. J. (Frère Gabriel-Marie). 1893. *Éxercices de Géométrie descriptive*. Tours: Alfred Mame et Fils París: Ch. Poussielgue.

FARISH, William. 1822. On Isometrical Perspective. *Cambridge Philosophical Society Transaction* I: 1-20.

FREZIER, Amédée-François. 1737-1739. *La théorie et la pratique de la coupe des pierres et des boï ... ou traité de stéréotomie a l'usage de l'architecture*. Strasbourg-Paris: Jean Danie Doulsseker-L. H. Guerin.

HACHETTE, Jean N. P. 1828. *Traité de Géométrie descriptive*. Paris: Corby (1st ed. 1822).

LA RUE, Jean Baptiste de. 1728. *Traité de la Coupe des pierres*. Paris: Imprimirie Royale.

LA GOURNERIE, Jules M. de. 1855. *Discours sur l'Art du trait et la Géometrie descriptive*, Parí Mallet-Bachelier.

——. 1858. *Traité de perspective linéaire*. París: Gauthier-Villars et fils.

——. 1860-64. *Traité de Géométrie descriptive*, 3 vols. Paris: Mallet-Bachelier.

——. 1867. *Recherches sur les surfaces réglées tétraédrales symétrique*. Paris: Gauthier-Villars.

——. 1872. *Mémoire sur l'appareil de l'arc biaise*. Paris: Libraire Polytechnique.

————. 1874. Mémoire sur l'enseignement des arts graphiques. *Journal des Mathématiques pures et appliqués* **XIX**: 113-156.

LAURENT, Roger. 1992. Théorie des ombres et des pénombres, perspective, perspective aérienne. pp. 547-563 in *L'École Normale de l'an III: Leçons de Mathématiques*, Jean Dhombres, ed. Paris: Dunot.

LORIA, Gino. 1921. *Storia della geometria descriptiva dalle origini sino ai giorni nostri*. Milan: Hoepli.

MONGE, Gaspard. 1796. Analyse appliqué à la Géométrie. Sur les lignes de courbure de la surface de l'Ellipsoïde. *Journal de l'École polytechnique* **2**: 145-165.

————. 1799 [An VII]. *Géométrie descriptive, leçons données aux Écoles normales, l'an 3 de la République ... .* Paris: Baudouin

————. 1820. *Géométrie descriptive*, Barnabé Brisson ed. Paris: Mme. Ve. Courcier.

————. 1801. *Feuilles d'Analyse appliqué à la Géométrie à l'usage de l'École Polytechnique*. Paris: Bernard (Reissued *as Application de l'Analyse à la Géométrie* in 1805, 1809, 1850).

OLIVIER, Théodore. 1847. *Applications de la Géométrie descriptive*. Paris: Carilian-Goeury et Vor. Dalmont.

————. 1852. *Cours de Géométrie descriptive*. Paris: Carilian-Goeury et Vor. Dalmont. (1st. ed. 1845)

PIRENNE, M.H. 1970. *Optics, Paintings & Photography*. London: Cambridge University Press.

PONCELET, Jean-Victor. 1822. *Traité des propriétés projectives des figures*. Paris: Bachelier.

POUDRA, Nöel-Germinal. 1859. *Examen critique du Traité de Perspective Lineaire de M. de La Gournerie*. Paris: J. Corréad.

————. 1864. *Histoire de la perspective ancienne et moderne*. Paris: J. Corréad

RABASA DÍAZ, Enrique. 1996. Arcos esviados y puentes oblicuos: el pretexto de la estereotomía del siglo XIX. *Obra pública* **38**: 30-41.

————. 1998. La bóveda plana de Abeille en Lugo. Pp. 409-415 in *Actas del segundo congreso nacional de historia de la construcción*. Madrid: Instituto Juan de Herrera.

————. 2000. *Forma y construcción en piedra: de la cantería medieval a la estereotomía del siglo XIX*. Madrid: Akal.

ROVIRA Y RABASSA, Antonio. 1897-1899. *Estereotomía de la piedra*, 2 vols. Barcelona: Estampería Artística.

SAKAROVITCH, Jöel. 1998. *Épures d'Architecture: De la coupe des pierres à la géométrie descriptive*. Basel: Birkhäuser.

TATON, René. 1951. *L'œuvre scientifique de Monge*. Paris: P.U.F.

VALLEE, Louis-Leger. 1819. *Traité de la géométrie descriptive*. Paris: Courcier.

*About the author*

Enrique Rabasa Díaz, born in 1957, holds a Ph.D. in Architecture and is Professor of the School of Architecture of the Technical University of Madrid (UPM). He belongs to the Architectural Graphic Design Department, and teaches in the grade and in the Master of Preservation of the UPM. His areas of research are history of descriptive geometry and history of construction, especially stereotomy and stonecutting. He manages a stonecutting workshop in the School, where the students go through the whole process of historic techniques of construction in stone: tracing, manual carving of stone and setting. He has written many articles, and is the author of the book *Forma y construcción en piedra: de la cantería medieval a la estereotomía del siglo XIX* (Madrid: Akal, 2000). He collaborates with the Centro de los Oficios of León, and is also the author of *Guía práctica de la estereotomía de la piedra* (León: Editorial de los Oficios, 2007), showing several actual stereotomic constructions made in this centre, its geometrical problems and how they were made step-by-step.

# Michael J. Ostwald

University of Newcastle
School of Architecture
and Built Environment
Faculty of Engineering
and Built Environment
Callaghan, New South Wales
2398 AUSTRALIA
michael.ostwald@newcastle.edu.au

Keywords: Glenn Murcutt,
Justified plan graph, graph theory,
Space Syntax, mathematical
analysis, plan analysis

This is the second part of a two-
part series. See: "The Mathematics
of Spatial Configuration:
Revisiting, Revising and Critiquing
Justified Plan Graph Theory",
*NNJ* 13, 2 (Summer 2011):
pp. 445-470.

Research

# A Justified Plan Graph Analysis of the Early Houses (1975-1982) of Glenn Murcutt

**Abstract.** The Justified Plan Graph (JPG) technique was developed in the late 1970s and refined in the following two decades as a means of undertaking qualitative and quantitative research into the spatial structure or permeability of buildings. Famously used by Space Syntax researchers to uncover the social logic of architectural types, the technique remains an important, if not widely understood, approach to the analysis of the built environment. This paper uses the JPG method to undertake a three-stage analysis of the early houses of Pritzker Prize winning architect Glenn Murcutt; the stages are visual analysis, mathematical analysis and theoretical analysis. Through this process the paper offers a rare application of the JPG method to multiple works by the same architect and demonstrates the construction of a series of "inequality genotypes", a partial "statistical genotype" and a partial "statistical archetype" for these houses. Instead of seeking to uncover the social structure of Murcutt's housing, the paper analyses the architect's distinctive approach to ordering space within otherwise simple volumes or forms. The ultimate purpose of this analysis is to offer an alternative space-based, rather than form-based, insight into this architect's work.

## Introduction

Architectural design analysis – the investigation of the properties, qualities and ideas found in a specific architect's work – remains almost exclusively focussed on questions of form and tectonics [Gelernter 1995; Frampton 1995; Baker 1996]. While stylistic, phenomenal and semiotic debates still occur about particular buildings, the canonical works of architectural history remain steadfastly focussed on formal properties, revisiting the key volumetric and material qualities of a building until, over time, a seemingly definitive reading of an architect's work has been reached. For example, the works of the Australian architect Glenn Murcutt have, over time, begun to be described in a highly consistent manner. With few exceptions, Murcutt's early rural domestic architecture has been delineated by historians as providing an exemplar of Arcadian minimalism – a rigorous modern evocation of the form and tectonics of the primitive hut. For example, Philip Drew proposes that Murcutt's talent lies in his the capacity to shape "a minimalism that is austere and tough so that all that remains is an irreducible core" [1986: 60]. Rory Spence describes Murcutt's early houses as constituting a clear formal type: "the long thin open pavilion" [1986: 72]. Francoise Fromonot argues that Murcutt's houses are all "variations on the same theme" and that these design "prototypes" represent a "relatively homogenous body of work. An analysis of [which] reveals a number of constants which could be called *characteristic*, analogous to those identifiable in specimens which illustrate a *species*" [2003: 60]. Drew, Fromonot and Spence are not alone in identifying in each of these houses a local variant of a more universal type. Yet, despite this apparent accord concerning formal qualities, relatively little has been said about Murcutt's architecture in terms of its spatial structure.

Space is the inverse of form [Ching 2007]; it is defined by walls and controlled by doors. A building form may also shape exterior space, but internal space is completely controlled by form [Unwin 2003]. Given this seemingly contingent relationship, it might be assumed that an analysis of spatial configuration in the architecture of Glenn Murcutt would simply support the more common formal reading. However, despite extensive critique of his buildings from formal, environmental, aesthetic or phenomenal perspectives, the spatial structure of the work remains only superficially described. For example, a number of passing references are made to Murcutt's plans featuring clear separation between served and service zones [Drew 1985, 2001; Beck and Cooper 2002; Frampton 2006]. Other than this general observation, Juhani Pallasmaa is the only critic to comment directly on spatial configuration, when he asserts that for Murcutt, order in form is as important as order in "organising and structuring" space [2006: 19]. He reinforces this point by proposing that Murcutt "doesn't merely aestheticise the human domicile"; he structures his designs to support "a humanised reading and meaning [of] the human condition itself" [2006: 17]. Pallasmaa's assertion is broadly that the rigour and simplicity of Murcutt's formal resolution is reflected in a similarly rigorous and refined spatial structure.

In response to the lack of spatial analysis to complement the existing, extensive formal analysis, the present paper uses the Justified Plan Graph (JPG) method to construct a graphical, mathematical and theoretical analysis of the spatial configuration of the first five of Murcutt's famous rural houses: the Marie Short House, the Nicholas House, the Carruthers House, the Fredericks Farmhouse and the Ball-Eastaway House. These houses were acknowledged in Murcutt's 2002 Pritzker Prize citation as being instrumental in shaping his international reputation. In much the same way that Fromonot [2003] describes these houses as specimens of the same species, in Space Syntax terms, they could be regarded as constituting important local *phenotypes* that represent singular variations of an overarching *genotype*.

While the theory and use of the JPG is well developed [Hillier and Hanson 1984; Hillier 1995], and stable computational versions of the method are available, there are relatively few examples of its application for the analysis of sets of architects' works. This paper seeks, in part, to revive the method through its application in two ways. First, the mathematical potential of the JPG has rarely been applied in longitudinal design analysis in this way and only a small number of precedents exist [Hanson 1998; Major and Sarris 1999; Bafna 1999]. Second, the paper proposes the construction of a simple statistical archetype from the various genotype examples in the longitudinal set. Notwithstanding these variations of the methodology, Murcutt's architecture has rarely been subjected to any form of mathematical or computational analysis in the past. The only exceptions to this include a shape grammar analysis of form undertaken by Hansen and Radford [1986a; 1986b].

In the following section a brief overview of Space Syntax and the JPG method is provided. Thereafter, the three major precedents to this study are considered in order to derive an approach for the research. Once this is outlined, then each of the five houses is described in chronological order, commencing with a traditional historical description before producing a JPG for visual analysis, mathematical analysis and then review. In the penultimate section the results of all five works are discussed together in the context of the set of inequality genotypes as a precursor to constructing a statistical archetype. Finally the conclusion contextualises the results and reflects on Pallasmaa's [2006] claims about the nature of Murcutt's planning.

While only a limited description of Space Syntax and the JPG method is included hereafter, the conceptual, mathematical, and theoretical background to the present paper appeared in the previous issue of the *Nexus Network Journal* (Ostwald 2011). That same paper also includes full worked examples of the method, a complete set of the formulas and an explanation of the nomenclature ( $i$, $TD$, $H^*$, etc.).

## The JPG method

Space Syntax promotes a conceptual shift in understanding architecture wherein "dimensional" or "geographic" thinking is rejected in favour of "relational" or "topological" reasoning [Hillier and Hanson 1984]. That is, the approach focuses on space, not form, and, more particularly, on non-dimensional qualities of space like permeability, control or hierarchy. This shift in thinking commences with the process of translating architecturally defined space into a series of topological graphs that may be visually inspected, mathematically analysed (graph *analysis*) and then interpreted (graph *theory*) in terms of their architectural, urban, social or spatial characteristics. While Space Syntax research has developed a wide range of possible methods for investigating the built environment, the present paper is only concerned with one approach; the JPG.

The first step in the construction of a JPG is typically the production of a convex map or boundary map. A convex map is a way of partitioning an architectural plan into a diagram of defined spaces or nodes and the connections between them. There are a number of alternative variations of this stage, ranging from the highly proscribed to the very general [Hillier and Hanson 1984; Markus 1993].

The particular method chosen for producing the convex map has a direct impact on the JPG and its results. For example, it is possible for an irregular plan for a small house to require as many as 50 separate convex spaces to fulfil the requirements of the original convex map definition [Hillier and Hanson 1984]. The subsequent JPGs are typically over-convoluted and can be mathematically dominated by the influence of often quite small architectural features. For example, the convex map produced by Major and Sarris [1999] of Peter Eisenman's House 1, has 39 nodes or spaces, while Eisenman identifies only seven functional spaces in the house! By counting every alcove for a built-in bookcase, display stand or wardrobe, and by dividing every section of space visually occluded by, or separated from, another space by a change in corridor width, or location of a blade column or open stair, the number of spaces can increase sixfold. This process artificially inflates the program and alters the actual, inhabited and experienced structure of the house. The more recent methods, as discussed in the next section, are more inclined to associate spaces directly with functional zones, thus reducing the number of nodes and more clearly aligning the JPG with inhabitation patterns [Peponis et al. 1997; Bafna 2003].

Once the convex plan is constructed it is converted into a graph diagram that displays only nodes (rooms) and lines (connections between rooms). This graph is arrayed across a number of levels, starting with zero at the base, regardless of the actual orientation of space in the original building [Hillier and Hanson 1984]. Once completed, the JPG displays levels of connectivity and separation between the root or carrier space, at the bottom of the JPG, and all other spaces. Thereafter, there are three common ways to approach the JPG.

First, a JPG may be graphically or visually analysed to uncover a range of qualitative properties of the spatial structure, including relative asymmetry, spatial hierarchy (arborescent qualities) and permeability (rhizomorphous qualities). The majority of the examples of this approach to the JPG are concerned with "inhabitant-visitor relations" and they rely on the production of JPGs with the exterior as carrier [Marcus 1987, 1993; Dovey 1999, 2010]. Despite this, a small number of examples of visual analysis have used multiple carriers and visual archetypes to investigate the properties of space [Alexander 1966; Ostwald 1997].

Second, the JPG may be mathematically analysed as a complete system. The formulas for this process may be found in a range of places [Hillier and Hanson 1984; Osman and Suliman 1994; Hanson 1998] as well as in several software tools (Depthmap; AGraph). From this analysis it is possible to develop a set of values describing the JPG from the point of view of Total Depth ($TD$), Mean Depth ($MD$), Relative Asymmetry ($RA$), integration ($i$) and control value ($CV$). $i$ values may be used in architectural analysis to develop an "inequality genotype", which is important in the present context because it formed the basis for the two major analytical precedents for the present paper [Major and Sarris 1999; Bafna 1999]. As Sonit Bafna [2001] explains,

> [t]he most common basis of comparison has been ... the inequality genotype: the ranking of programmatic labeled spaces according to their mean depth (most often described in terms of *integration* values) of the nodes in the graph of the spatial configuration to which they correspond [Bafna 2001: 20.1].

In practice, an inequality genotype is a list of spaces in the JPG, arranged in order from highest to lowest $i$ value. But in order to interpret what this list means, we have to leave behind the mathematics and start to consider wider social and cultural factors that are part of graph theory.

The visual and mathematical information derived from the JPG may be used to theorise some additional properties or qualities about a building. This, the third approach, is the most controversial [Dovey 1999] but it is also necessary for any attempt to use the JPG to assist in interpreting architecture. For example, returning to the inequality genotype, Zako argues that it is "one of the most general means by which culture is built into spatial layout" [2006: 67]. However, the inequality genotype is simply a hierarchical list, and to interpret further how deliberate it is, it must be interpreted with the assistance of the difference factor ($H$). Zako notes that the *difference factor* "was developed to quantify the degree of difference between the integration value of any three (or more with a modified formula) spaces or functions" [2006: 67]. Therefore, the difference factor, or $H$, can be used to determine how strong or weak certain inequalities are in the base JPG. Thus, an inequality genotype with "a low entropy [$H$] value will therefore be [a] 'strong' genotype, whereas one that exists, but tends to have a high entropy, will be a 'weak' genotype" [Zako 2006: 67]. This is a typical example of a reasonably accepted use of mathematics to hypothesize certain qualities about an architectural plan.

A less emphatic interpretation is offered by Hillier and Tzortzi, who propose that through the application of visual and mathematical processes, a JPG can be used to demonstrate how a "culture manifests itself in the layout of space by forming a spatial pattern in which activities are integrated and segregated to different degrees" [2006: 285]. This is possible because the spaces are not just multi-purpose voids awaiting appropriate furnishings and fittings, but they are also locked into a "certain configurational relation to the house as a whole" [2006: 285]. It is for this reason that the inequality genotype is used to uncover not only a set of social values or ideals responsible for shaping architecture, but also the recurring social values and principles in an individual architect's works.

## Methodological precedents

Hanson's study of housing [1998], using a combination of JPG and axial graph methods, includes a qualitative review of the plans of several famous houses by Adolf Loos, John Hejduk, Mario Botta and Richard Meier. Without the support of a mathematical analysis, Hanson's review of these houses is largely restricted to identifying differences in the visual structure of the JPGs. While this is an important early example, it tends to be of limited practical use in the present context because it relies on a rigid formula for convex map construction (leading to over 60 defined nodes in several cases). The following year, Major and Sarris set out to use the JPG method to analyse eight houses by Peter Eisenman [1999]. In each case they produced a JPG for visual analysis and then used mathematics to develop an inequality genotype recording the order of integration of spaces from highest to lowest. One of the important issues in Major and Sarris's work which is relevant for the present research is that, by using the original method of convex plan generation, they produced JPGs for houses with up to 133 separate spaces. As a result of this process, their inequality genotypes did not display a high level of order until they had been stripped of all but the major functional zones.

Probably the best precedent for the present study is found in the work of Bafna [1999; 2001] who has published several JPG analyses of Mies van der Rohe courtyard houses. What is significant in Bafna's work is that he too has found that inequality genotypes are difficult to work with. Bafna's not unreasonable starting assumption was that there would be a "genotypical consistency in these houses" which could be used "as a basis upon which to study their phenotypical differences" [2001: 20.3]. This implies that the order of rooms in the inequality genotype would reflect the architect's ideal (itself a reflection of social conditions) and that small differences in the JPG would be the result of differences caused by particular, site, context of program conditions. Unfortunately, the inequality genotypes were more diverse than anticipated, and even simplifying the node set (as Major and Sarris were forced to do) did not produce a clear result. Upon reflection, this realisation lead Bafna [2001] to conclude that the genotype is "better defined, not as a given rank order of labeled spaces, but [as] a statistically stable pattern of variation of those" [2001: 20.9]. Thus, it is the broader pattern represented in the genotypes that is most important.

As described in the following section, the present paper adopts a variation of Bafna's [1999, 2001] convex map boundary generation rules and the inequality genotype method. Rather than copying Bafna's mathematical approach to the genotype analysis, the present paper uses a simplified statistical version to identify both genotype patterns and a stable archetype in Murcutt's early house designs.

## Approach

Despite completing several urban houses prior to 1975, the five houses being analysed in the present paper are widely regarded as the first of Murcutt's characteristic works. Drew describes the first four of these houses as a significant set; the "Marie Short, Nicholas, Carruthers and Fredericks farmhouses are really members of a series, … taken together, they represent a progressive development and refinement of the longitudinal house type" [1985: 92]. These four also directly prefigure a fifth house – an intermediate work in Murcutt's oeuvre –, the Ball-Eastaway House [Farrelly 1993; Fromonot 1995]. After the completion of the Ball-Eastaway house, Murcutt retained his linear planning style but he developed more elaborate sections, typically featuring curvilinear steel structures, as well as producing a series of larger houses. Significantly, all of the five houses considered in this paper have been altered or extended since being completed and many have been resold (Murcutt himself now owns the Marie Short House). In all cases, the version of the house analysed here is the original, and the original naming of each house has also been retained. Furthermore, several of the houses feature mezzanine levels that are rarely acknowledged in published plans or sections, and many of these are not even apparent in published photographs. In the present paper the mezzanine levels that were completed as part of the original construction phase are all included in the analysis.

As the first stage in the process, new plans for each house were prepared and annotated with a standard set of abbreviations and in accordance with Murcutt's original notations (table 1). The general principle adopted in this paper for the construction of a convex map or boundary map, is to try to keep the set as small or economical as possible. Thus, the method broadly follows Bafna's approach:

> [The JPGs are based] upon a modified version of the boundary map of the plans, rather than go with the more conventional minimum convex partition. One reason for this was that minimum convex partition generates spaces to which programmatic labels might be difficult to assign; another, that it is based upon a heuristic method which, given the free-plan arrangement of several houses, could be quite inconsistent. The boundary map, by contrast, is generated by recognizing programmatically defined boundaries between individual components [Bafna 1999: 01.7]

This approach is ideal for analyzing the work of Murcutt, an architect who often designs small alcoves to accommodate cupboard door swings or ledges to display artwork. None of these alcoves, ledges or indented walls are separately identified. Furthermore many of Murcutt's houses, being rural in their settings, have small utility zones as their secondary connection to the exterior. Such zones typically feature storage cupboards, hanging rails (for coats) and racks for shoes. In this paper utility zones are typically identified, according to Murcutt's labels, as a single area even though a rigorous convex mapping exercise would divide them into as many as nine separate spaces. Similarly bathrooms with internal, partial-height partitions separating bath, shower or toilet are counted as one space.

The issue of open plan space is more complex, with many of Murcutt's spaces being defined by a combination of the furniture in them and the label on a plan. If we follow Bafna [1999], then some of Murcutt's major spaces, which have no visible separation but are labelled dining room, living room or kitchen, would be divided along these lines. However, in the present paper, a larger set of threshold conditions has been required to break down an open plan space. For example, a change in floor texture alone is no enough to signal a new space, but in combination with a freestanding column or an island kitchen bench, it can identify a separate zone, even if it remains otherwise almost completely within an open-plan volume. Where no clear threshold marker combinations existed, even though multiple functions were labelled in the one space, the space was counted as a singular node. Because several of Murcutt's houses feature open-plan, multifunction areas with no clear thresholds, some combined space labels were used (for example, KDL, means an open plan, kitchen, dining and living space).

While not strictly relevant to the method, to assist the reader large open plan spaces have elongated oval nodes as opposed to circular ones in the JPG. With the exception of the exterior, if there are more than one of a particular room type they are numbered; thus if there are two bedrooms, they become B1 and B2. The exterior is represented in the JPG as a crossed circle and in text and tables as ⊕. Three additional graphic conventions were adopted to assist the reader. First, a double line in a JPG indicates a primary car access, next, a line in a JPG broken by a zigzag indicates a major change in level (typically stairs or a ladder) and third, a dashed and dotted line has been used for secondary or service access from the exterior. None of the graphic variations described in this paragraph have an impact on the mathematical results. Similarly, JPG theory uses curved lines to connect nodes when required by the visual complexity of the graph. Curved lines are effectively identical to straight lines in a graph as far as the mathematical analysis is concerned.

| ⊕ | Exterior | L | Living Area | B | Bedroom |
|---|---|---|---|---|---|
| V | Veranda | F | Family Room | WIR | Walk in Robe |
| LBY | Lobby | D | Dining | b | Bathroom |
| C | Court | K | Kitchen | SH | Shower |
| H | Hall | ST | Studio | WC | Toilet |
| G | Garage | M | Music Room | A | Alcove |
| U | Utility Room | l | Laundry | S | Store |

Table 1. General abbreviations for rooms used in the JPG

## Five Houses

### Marie Short House

The Marie Short House (1975) is sited on a raised floodplain, in the bend of a river, near Kempsey, in northern New South Wales, Australia. This is the first of Murcutt's famous regional houses; it was credited as heralding both a new Australian style [MacMahon 2001] as well as being a key Critical Regionalist work [Frampton 2006]. The house consists of two, similarly sized pavilions that are placed side-by-side and then slid apart along a centreline. One pavilion contains living spaces, the second, sleeping quarters, and a corridor both divides and connects the two (figs. 1, 2). Drew describes the house as featuring a pair of pavilions which merge "Mies van der Rohe's single storey glass pavilion type" with the "primitive hut archetype" [1985: 74]. Beck and Cooper note that the staggered plan of the Marie Short House "is reminiscent of the Farnsworth house with its staggered deck" and "layered zones of public and private" [2002: 48]. Frampton reiterates this canonical reading of the house as capturing the essence of the "Semperian primitive hut of 1852 with the tectonic refinement of Mies' Farnsworth House" [Frampton 2002: 1].

A visual analysis of the JPG for the Marie Short House (fig. 3) reveals an unexpectedly complex, two-part structure, with a "ring-like" circulation approach to the living pavilion and a "bush-like" structure, rooted in the hallway, to the more private pavilion, Between these two, the hall provides a constant point of connection and passage. While far more complex and "deep" than might be imagined from its simple exterior form, this spatial pattern is reasonable for two adjacent structures, one of which is more flexible for living and the other of which is compartmentalised for private activities (sleeping and bathing).

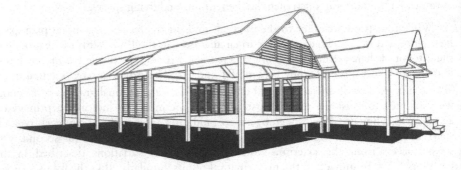

Fig. 1. Perspective, the Marie Short House (1975), Glenn Murcutt

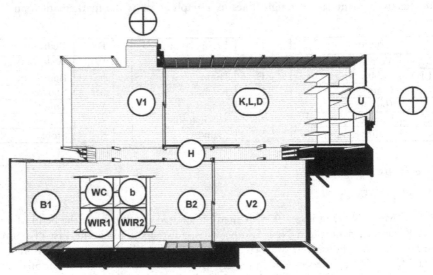

Fig. 2. Annotated Plan, the Marie Short House (refer to Table 1 for room abbreviations)

The mathematical results for the Marie Short JPG (table 2) show that the mean total depth ($TD$) of the house is 24.83. Conversely, the mean depth ($MD$) of rooms in the house is 2.25; this suggests that the most isolated spaces in the configuration are, in order, the two walk-in-wardrobes ($MD = 3.00$ and $MD = 2.90$), the utility are ($MD = 2.63$) and the exterior ($MD = 2.63$). Conversely, the most accessible are the hallway ($MD = 1.36$), veranda 1 ($MD = 1.90$) and the open plan kitchen dining and living areas ($MD = 1.90$). The integration ($i$) values confirm this, but provide advice on the relative magnitude of the integration or isolation. For example, the hallway ($i$ 13.75) is more than double the level of the next most integrated pair of spaces, veranda and the kitchen, living and dining areas (both, $i = 5.50$). The remainder of the rooms including the exterior, are all relatively isolated (ranging from 3.00 to 5.00) with only the two wardrobes being markedly isolated. Finally, and not surprisingly, the hall exerts the highest spatial influence ($CV = 4.00$) generally possessing more than four times the capacity to influence space of any of the other nodes. The only anomaly in the analysis is that bedroom two, with its access to a second private veranda, is also surprisingly highly integrated ($i = 5.00$), and exerts the second highest level of control in the house ($CV = 1.64$).

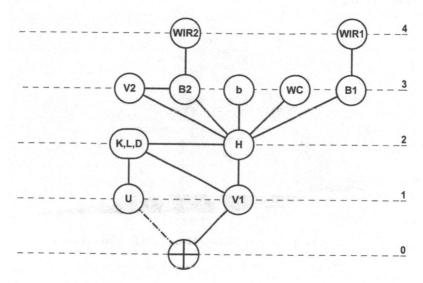

Fig. 3. JPG, with exterior as carrier, for the Marie Short House

| # | Space | $TD_n$ | $MD_n$ | RA | i | CV |
|---|-------|--------|--------|-----|------|------|
| 0 | ⊕ | 29 | 2.63 | 0.32 | 3.05 | 0.83 |
| 1 | V1 | 21 | 1.90 | 0.18 | 5.50 | 0.97 |
| 2 | K,L,D | 21 | 1.90 | 0.18 | 5.50 | 0.97 |
| 3 | H | 15 | 1.36 | 0.07 | 13.75 | 4.00 |
| 4 | U | 29 | 2.63 | 0.32 | 3.05 | 0.83 |
| 5 | V2 | 23 | 2.09 | 0.21 | 4.58 | 0.47 |
| 6 | B2 | 22 | 2.00 | 0.20 | 5.00 | 1.64 |
| 7 | B | 25 | 2.27 | 0.25 | 3.92 | 0.14 |
| 8 | WC | 25 | 2.27 | 0.25 | 3.92 | 0.14 |
| 9 | B1 | 23 | 2.09 | 0.21 | 4.58 | 1.14 |
| 10 | WIR2 | 32 | 2.90 | 0.38 | 2.61 | 0.33 |
| 11 | WIR1 | 33 | 3.00 | 0.40 | 2.50 | 0.50 |
| Minimum | | 15.00 | 1.36 | 0.07 | 2.50 | 0.14 |
| Mean | | 24.83 | 2.25 | 0.25 | 4.83 | 1.00 |
| Maximum | | 33.00 | 3.00 | 0.40 | 13.75 | 4.00 |
| H | | 0.92 | | H* | | 0.56 |

Table 2. Summary of JPG results for the Marie Short House

## Nicholas House

Located in the Blue Mountains, west of Sydney, the Nicholas House (1980) and the Carruthers House (1980) – discussed in the next section – were built on adjacent sites as country retreats for the families of two lawyers. While the Nicholas House, like the Marie Short house, has a two-pavilion *parti*, it is the first of Murcutt's houses where the pavilions are unequally sized to accommodate living spaces in the larger one and services in the smaller (fig. 4).

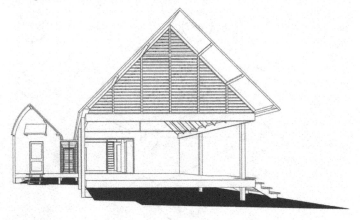

Fig. 4. Perspective, the Nicholas House (1980), Glenn Murcutt

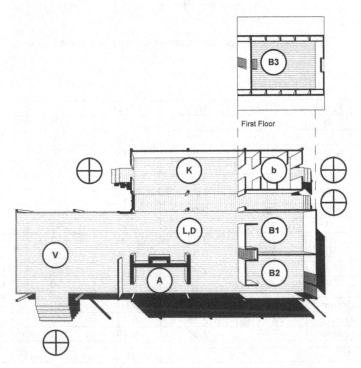

Fig. 5. Annotated Plan for the Nicholas House

The larger north pavilion of the Nicholas house is dominated by semi-open plan living and eating areas as well as two ground floor bedrooms. A loft space, accessed by a narrow ladder, is created for the third bedroom. This main pavilion, which like most of Murcutt's houses is slightly raised above the ground, is clad in timber boards and lined with glass louvers and cedar external blinds. In contrast, the south edge of the house has a distinctive solid wall clad in corrugated iron and with a curved roof above. The service zones, including the kitchen, bathroom and storage, are located in this smaller pavilion (fig. 5).

A visual review of the JPG for the Nicholas house reveals a shallow structure that is three levels deep for the exterior carrier (the Marie Short house was four levels deep for the same carrier), with a "ring-like" entry configuration encompassing the exterior, veranda, living and dining and kitchen. The living and dining spaces are the starting point for a "bush-like" structure extending beyond that for the more private areas (fig. 6). Though of a slightly smaller scale than the Marie Short House (eight spaces as opposed to eleven), and lacking an explicit hallway connection, the configurational strategy is unexpectedly similar.

The mathematical analysis of the Nicholas house reveals that while it is only marginally smaller in program than the Marie Short House, it is much simpler in its configuration (table 3). The average total depth of the Nicholas house plan is 15.55, which is around 60% less than the result for the Marie Short House. The most integrated space in the Nicholas house is the combined living and dining room, which connects the majority of the plan. For the remainder of the spaces, the next most integrated is the kitchen ($i$ = 5.60), closely followed by the veranda ($i$ = 4.66) and a range of spaces thereafter with equal integration vales ($i$ = 3.11).

The open plan living and dining area exerts the greatest degree of control, with a $CV$ value of 4.83; a result which is almost five times higher than the nearest spatial competitor — the kitchen — and more than ten times higher than for the majority of the rooms. This not only confirms the intuitive reading that the open plan living and dining area is the most important in the house, it quantifies the level of importance of that space.

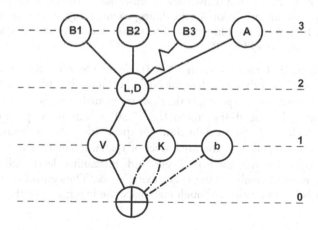

Fig. 6: JPG, with exterior as carrier, for the Nicholas House

| # | Space | $TD_n$ | $MD_n$ | RA | i | CV |
|---|---|---|---|---|---|---|
| 0 | ⊕ | 17 | 2.12 | 0.32 | 3.11 | 1.33 |
| 1 | V | 14 | 1.75 | 0.21 | 4.66 | 0.50 |
| 2 | K | 13 | 1.62 | 0.17 | 5.60 | 1.00 |
| 3 | b | 18 | 2.25 | 0.35 | 2.80 | 0.66 |
| 4 | LD | 10 | 1.25 | 0.07 | 14.00 | 4.83 |
| 5 | B1 | 17 | 2.12 | 0.32 | 3.11 | 0.16 |
| 6 | B2 | 17 | 2.12 | 0.32 | 3.11 | 0.16 |
| 7 | B3 | 17 | 2.12 | 0.32 | 3.11 | 0.16 |
| 8 | A | 17 | 2.12 | 0.32 | 3.11 | 0.16 |
| Minimum | | 10.00 | 1.25 | 0.07 | 2.80 | 0.16 |
| Mean | | *15.55* | *1.94* | *0.26* | 4.73 | 1.00 |
| Maximum | | 18.00 | 2.25 | 0.35 | 14.00 | 4.83 |
| H | | 0.94 | | H* | | 0.61 |

Table 3: Summary of JPG results for the Nicholas House

## Carruthers House

Located on the site adjacent to the Nicholas House, the Carruthers House (1980) is, at first glance, even more straightforward in its form and design. Fromonot describes it as a "simple timber barn roofed with corrugated iron" [2003: 112]. With the exception of the chimney, the single pavilion sits lightly on posts above the ground plane. Internally it is divided into two sections, the north edge that contains the main circulation space and a sitting room open to the landscape and the south edge where bedrooms, a bathroom and a kitchen are located. At one end of the pavilion there is a loft bedroom, while at the other the living area has a large, double height space. Externally, the south wall is almost fully enclosed protecting the inhabitants from winter winds. The actual building contains four elevated water collection tanks, which change its character, but otherwise have no impact on the present study (fig. 7).

A visual analysis of the JPG for the Carruthers House reveals a shallow structure (three levels of depth), with a dense, nested 'bush-like" structure with primary "root" in the hallway and secondary "root" in the dining room (really an extension of the hallway spatially, but because of the placement of furniture and the mezzanine above, a distinct and separate spatial zone) (fig. 9).

The mathematical analysis of the JPG (table 4) identifies the mean depth of the structure as 16.88; slightly more than the Nicholas House but still less complex than the Marie Short House. Not surprisingly the most integrated space is the hallway ($i = 9.33$) closely following by the dining room ($i = 7.00$) which is, as previously stated, an extension of the hallway, then a tight cluster of spaces (the exterior, bedroom 1, bedroom 2 and the bathroom) all with the same level of spatial integration ($i = 2.80$). The least integrated spaces are the kitchen, living and mezzanine levels (all $i = 2.54$); an unexpectedly isolated result for the major living space. The control value results mostly reflect the integration results, although they place the living area in the middle group of results.

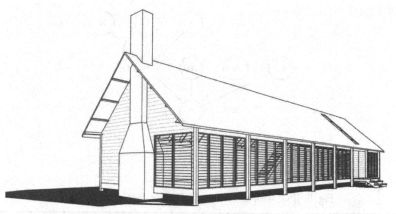

Fig. 7. Perspective, the Carruthers House (1980), Glenn Murcutt

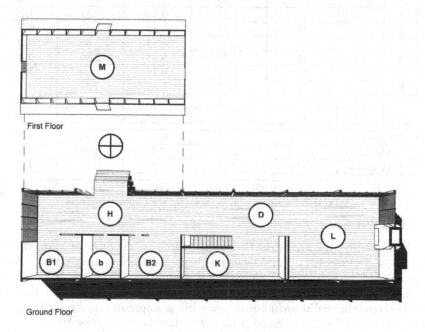

Fig. 8. Annotated Plan for the Carruthers House

There is an isolated phenomenal account of visiting this house that reflects some of these mathematical results. Drew argues that upon entry into the house the visitor is drawn into the "the pine tube" of the primary volume which is interrupted by three inserted planes: "one which separates the living room from the kitchen … one on the left of the stair, and another, below the left floor deck in line with the bedrooms, run parallel with the axis of the pavilion" [1985: 96]. The impact of these three spatial dividers is to lead the visitor to the sense that the space is "surge[ing] back and forth like a stream encountering boulders in its course" [1985: 96]. The two control values for the hallway and the dining room suggest a strong linear "pull" along the façade of the building that is interrupted by a series of side rooms, some irregularly placed, with lower control values.

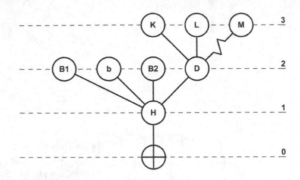

Fig. 9. JPG, with exterior as carrier, for the Carruthers House

| # | Space | $TD_n$ | $MD_n$ | RA | i | CV |
|---|-------|--------|--------|-----|------|------|
| 0 | ⊕ | 18 | 2.25 | 0.35 | 2.80 | 0.20 |
| 1 | H | 11 | 1.37 | 0.10 | 9.33 | 4.25 |
| 2 | B1 | 18 | 2.25 | 0.35 | 2.80 | 0.20 |
| 3 | b | 18 | 2.25 | 0.35 | 2.80 | 0.20 |
| 4 | B2 | 18 | 2.25 | 0.35 | 2.80 | 0.20 |
| 5 | D | 12 | 1.50 | 0.14 | 7.00 | 3.20 |
| 6 | K | 19 | 2.37 | 0.39 | 2.54 | 0.25 |
| 7 | L | 19 | 2.37 | 0.39 | 2.54 | 0.25 |
| 8 | M | 19 | 2.37 | 0.39 | 2.54 | 0.25 |
| Minimum | | 11.00 | 1.37 | 0.10 | 2.54 | 0.20 |
| Mean | | 16.88 | 2.11 | 0.31 | 3.90 | 1.00 |
| Maximum | | 19.00 | 2.37 | 0.39 | 9.33 | 4.25 |
| H | | 0.978 | | H* | | 0.703 |

Table 4. Summary of JPG results for the Carruthers House

## Fredericks Farmhouse

Drew describes the Fredericks Farmhouse (1982) as "the finest of Murcutt's series of long houses" [1985:121]. For Drew, this house achieves a relationship between the landscape and the form of the building that is reminiscent of a temple: "Classical without sacrificing any of its richness to oversimplification, light in appearance, it is the best kind of essentialist minimalist architecture, every bit as impressive as the landscape" [Drew 1985, 121]. The Fredericks house is located in Jambaroo, south of Sydney and slightly inland from the coast. Superficially, it appears to have a cross-section that is reminiscent of the Marie Short House, but in this case, while the two pavilions might have similar sections, they are very different in floor area (fig. 10). Both pavilions are timber, post and beam structures, with external western red cedar cladding. Murcutt describes the house as having "a very ordinary plan [...] like a railway carriage" [in Beck and Cooper 2002: 77]. An existing chimney structure anchors one side of the plan, with its central kitchen, dining and living spaces, while at each of the two ends of the pavilion there are bedroom, bathroom and services (fig. 11). Furthermore, this house has two loft-bedrooms which are rarely depicted in images or plans; Beck and Cooper argue that Murcutt's reluctance to introduce a loft space may be due to the "dynamic spatial condition that disturbs the serenity" of the rest of the house [2002: 76].

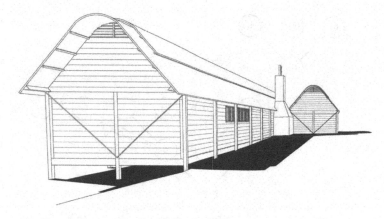

Fig. 10. Perspective, the Fredericks House (1980), Glenn Murcutt

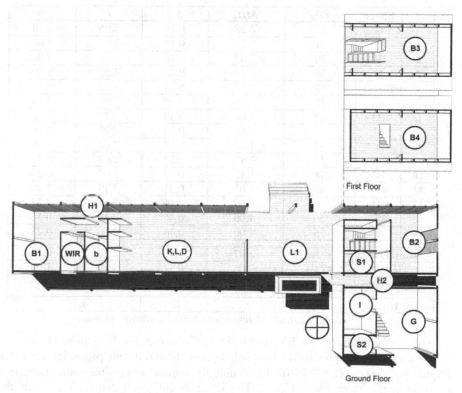

Fig. 11. Annotated Plan for the Fredericks House

A visual inspection of the JPG of the Fredericks House reveals a "ring-like" entry structure, leading to a primary "bush-like" private zone and a secondary "bush-like" private zone, along with some isolated service rooms (fig 12). This is the third of Murcutt's houses to feature a combination of "ring-like" and "bush-like" configurations, the second "bush-like" growth appears to be in response to the need to increase the number of spaces in the plan.

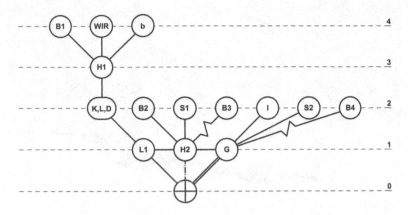

Fig. 12. JPG, with exterior as carrier, for the Fredericks House

| # | Space | $TD_n$ | $MD_n$ | RA | i | CV |
|---|-------|--------|--------|------|------|------|
| 0 | ⊕ | 32 | 2.28 | 0.19 | 5.05 | 0.70 |
| 1 | L1 | 31 | 2.21 | 0.18 | 5.35 | 1.00 |
| 2 | H2 | 29 | 2.07 | 0.16 | 6.06 | 3.86 |
| 3 | G | 35 | 2.50 | 0.23 | 4.33 | 3.50 |
| 4 | K,L,D | 36 | 2.57 | 0.24 | 4.13 | 0.58 |
| 5 | B2 | 42 | 3.00 | 0.30 | 3.25 | 0.16 |
| 6 | B3 | 42 | 3.00 | 0.30 | 3.25 | 0.16 |
| 7 | S1 | 42 | 3.00 | 0.30 | 3.25 | 0.16 |
| 8 | I | 48 | 3.42 | 0.37 | 2.67 | 0.20 |
| 9 | S2 | 48 | 3.42 | 0.37 | 2.67 | 0.20 |
| 10 | B4 | 48 | 3.42 | 0.37 | 2.67 | 0.20 |
| 11 | H1 | 43 | 3.07 | 0.31 | 3.13 | 3.50 |
| 12 | B1 | 56 | 4.00 | 0.46 | 2.16 | 0.25 |
| 13 | WIR | 56 | 4.00 | 0.46 | 2.16 | 0.25 |
| 14 | b | 56 | 4.00 | 0.46 | 2.16 | 0.25 |
| Minimum | | 29.00 | 2.07 | 0.16 | 2.16 | 0.16 |
| Mean | | 42.93 | 3.06 | 0.31 | 3.49 | 1.00 |
| Maximum | | 56.00 | 4.00 | 0.46 | 6.06 | 3.86 |
| H | | | 1.017 | | H* | 0.799 |

Table 5. Summary of JPG results for the Fredericks House

The mathematics of the JPG (table 5) confirms that the Fredericks House is the largest and most complex of the five early houses covered in this paper. Its mean Total Depth (mean $TD$ = 42.93) is roughly double the number of possible connections in the structure as the Marie Short House. This is significant, given that there are only three more spaces in the Fredericks House than in the Marie Short House. Paradoxically, the most integrated space is hallway 2 ($i$ = 6.06), followed by the living room ($i$ = 5.35), the exterior ($i$ = 5.05) and the garage ($i$ = 4.33); this is a mixed result with a service hallway and the garage featuring unusually strongly in the configuration. The least integrated spaces are more consistent with Murcutt's other works. These include bedroom 1 bathroom and the walk-in wardrobe ($i$ = 2.16). The control value results further crystallise this unexpected structure, with the most significant spaces being hallway 2 ($CV$ = 3.86), hallway 1 ($CV$ = 3.50) and the garage ($CV$ = 3.50)! The mathematical analysis suggests that the more complex the house, the more likely it is to rely on secondary circulation and spaces to achieve connectivity, and that this type of planning draws the user or visitor away from the major spaces in the house.

## Ball-Eastaway House

Designed as a house and private gallery for the artists Syd Ball and Lyn Eastaway, the Ball-Eastaway House (1982) is sited in top of a series of sandstone ledges near a wooded reserve to the northwest of Sydney. The Ball-Eastaway House has a "train carriage" plan with "a simple arrangement of rooms located beneath the gentle barrel-vaulted ceiling" [MacMahon 2001: 122]. The train carriage feeling is exaggerated externally with the building sitting above the ground, as if raised on wheels, and being clad in corrugated steel, with exposed downpipes and vents (fig. 13). Elizabeth Farrelly describes the carriage or pavilion form as being "[o]pen at both ends" leading the house to "became an extruded form [and] emphatically directional" [1993, 21]. Whereas the first four houses in the present set are clad largely in timber, and with exposed timber detailing, the Ball-Eastaway house has metal cladding, a more industrial feel, and is lined internally in white plasterboard.

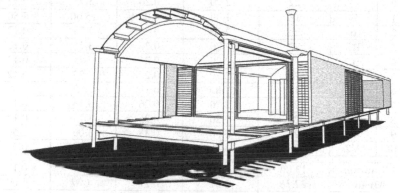

Fig. 13. Perspective, the Ball-Eastaway (1980), Glenn Murcutt

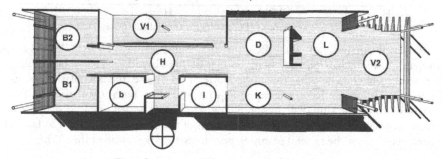

Fig. 14. Annotated Plan for the Ball-Eastaway

While this building appears to be a departure from Murcutt's previous aesthetic and tectonic practices, in planning terms it is closely associated with the previous four designs (fig. 14). Furthermore, despite often being left out of recent publications on Murcutt's work [Gusheh et al. 2008] – perhaps because it is not a clear example of critical regionalism – Fromonot describes the Ball-Eastaway house as "one of Murcutt's most successful buildings. It epitomises the lightweight, linear, economical and elegant pavilion, minimal in its environmental impact" [1995: 84].

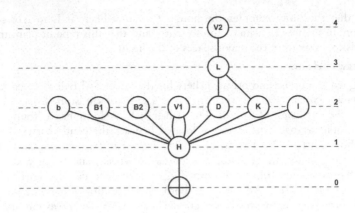

Fig. 15. JPG, with exterior as carrier, for the Ball-Eastaway

| # | Space | $TD_n$ | $MD_n$ | RA | i | CV |
|---|-------|--------|--------|-----|-----|-----|
| 0 | ⊕ | 22 | 2.20 | 0.26 | 3.75 | 0.12 |
| 1 | H | 13 | 1.30 | 0.06 | 15.00 | 7.50 |
| 2 | b | 22 | 2.20 | 0.26 | 3.75 | 0.12 |
| 3 | B1 | 22 | 2.20 | 0.26 | 3.75 | 0.12 |
| 4 | B2 | 22 | 2.20 | 0.26 | 3.75 | 0.12 |
| 5 | V1 | 22 | 2.20 | 0.26 | 3.75 | 0.12 |
| 6 | D | 18 | 1.80 | 0.17 | 5.62 | 0.62 |
| 7 | K | 22 | 2.20 | 0.26 | 3.75 | 0.12 |
| 8 | 1 | 22 | 2.20 | 0.26 | 3.75 | 0.12 |
| 9 | L | 25 | 2.50 | 0.33 | 3.00 | 1.50 |
| 10 | V2 | 34 | 3.40 | 0.53 | 1.87 | 0.50 |
| Minimum | | 13.00 | 1.30 | 0.06 | 1.87 | 0.12 |
| Mean | | 22.18 | 2.21 | 0.27 | 4.70 | 1.00 |
| Maximum | | 34.00 | 3.40 | 0.53 | 15.00 | 7.50 |
| H | | 0.848 | | H* | | 0.382 |

Table 6. Summary of JPG results for the Ball-Eastaway House

A visual analysis of the JPG for the Ball-Eastaway House shows a spatial configuration which is partway between that of the Carruthers House (a simple "bush-like" structure) and the other three; the Marie Short House, the Nicholas House and the Fredericks House all have a compound ring and then bush" structure. In the Ball-Eastaway House the ring (hall, living, dining, kitchen) is nested one level deep within the greater arborescent structure; a partial inversion of the fine grained pattern so far, but also a reinforcement of the general planning principles already identified (fig. 15).

The Ball-Eastaway House has a mean structural depth of 22.18 which is similar to that of the Marie Short House; both also have a similar number of rooms (table 6). Just as the JPG diagram for the exterior carrier implies, the hall is the most important room on the spatial configuration. It has an integration value of 15 and a control value of 7.50. This single hall, more than any individual room (or compartment) is the most important space in the everyday use of the house. Beyond the hall, veranda 1, the kitchen and the dining area, the remainder of the spaces are isolated and controlled by the hallway/room structure.

## Discussion

As the first step in attempting to identify the primary spatial patterns in Murcutt's early houses the inequality genotypes for each are recorded. These are as follows:

- **Marie Short House:**

  H (13.75) > V1 (5.50) = KLD (5.50) > B2 (5.00) > V2 (4.58) = B1 (4.58) > b (3.92) = WC (3.92) > ⊕ (3.05) = U (3.05) > WIR2 (2.61) > WIR1 (2.50).

- **Nicholas House:**

  LD (14) > K(5.60) > V (4.66) > ⊕ (3.11) = B1 (3.11) = B2 (3.11) = B3 (3.11) = A (3.11) > b (2.80).

- **Carruthers House:**

  H (9.33) > D (7.00) > ⊕ (2.80) = B1 (2.80) = B2 (2.80) = b (2.80) > K (2.54) = L (2.54) = M (2.54).

- **Fredericks Farmhouse:**

  H2 (6.06) > L1 (5.35) > ⊕ (5.05) > G (4.33) > KLD (4.13) > B2 (3.25) = B3 (3.25) = S1 (3.25) > H1 (3.13) > 1 (2.67) = S2 (2.67) = B4 (2.67) > B1 (2.16) = b (2.16) = WIR (2.16).

- **Ball-Eastaway House:**

  H (15) > D (5.62) > ⊕ (3.75) = b (3.75) = B1 (3.75) = B2 (3.75) = V1(3.75) = K (3.75) = 1 (3.75) > L (3.0) > V2 (1.87).

In four of these cases hallways dominate the genotype as the most integrated spaces, and in three of these cases bedrooms, bathrooms and walk-in-robes are the least integrated. As Bafna records, "[i]t is natural that the circulation areas and lobby will be more integrated and that the bedrooms and the services will occupy the other pole along the integration-segregation axis" [2001: 20.8]. But beyond these general tendencies, there is also unexpected variation in the inequality genotypes and particularly in regard to the more 'public' spaces, like the living areas, dining room or kitchen. For example, in the Carruthers House and the Ball-Eastaway House, the living areas are the second most isolated. In the other three houses, the living areas are amongst the most integrated.

In order to seek a clear pattern in the work, both Bafna [1999] and Major and Sarris [1999] simplified their inequality genotypes by removing singular room types and combining other similar sets of rooms so that the focus was on a smaller set of rooms which were present in all cases. In the present paper the following steps have been taken to achieve this:

1. Room types with less then three instances have been removed; that is, across the five houses, one alcove, one utility room, one mezzanine, a second veranda, two storerooms and two third bedrooms have been removed;

2. Hallways and verandas that function as circulation are grouped into one category and counted as the higher *i* value of the pair;

3. Bathrooms and toilets are grouped into one category and counted as the higher *i* value of the pair.

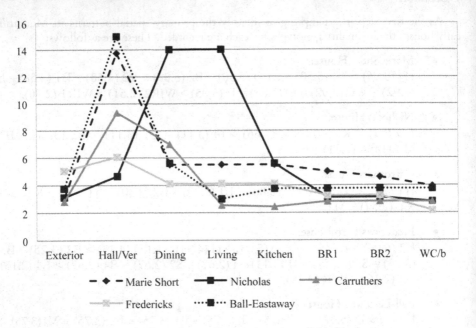

Fig. 16. Chart of inequality genotype data; divided by house

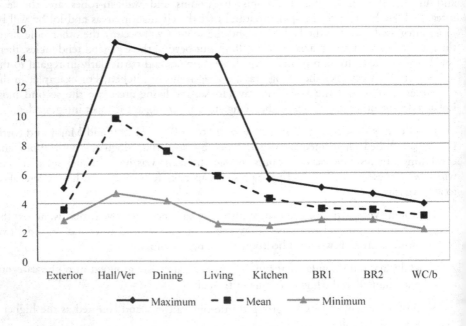

Fig. 17. Chart of inequality genotype data; divided by maximum, minimum and mean

Once these changes have been made, a simple pattern begins to emerge from the inequality genotypes, although there are still inconsistencies. In the comparative chart (fig. 16) the functional spaces are arranged along the $x$-axis broadly in accordance with the principal of "intimacy gradients" [Alexander et al. 1977]; that is, with the most public to the left and the least public to the right. The $y$-axis simply records the integration value. An intuitive reading of the relationship between integration and privacy might anticipate a line that commences at its highest level to the left of the $x$-axis and then drops to a low level at the right side. That is, the more public the space, the more integrated it is; the more private the space, the less integrated. The results of the graph are not quite this straightforward although, with the exception of the exterior, there is a broad trend down across the results from left to right, albeit typically across a series of plateaus. When the data is re-sorted to identify mean values, maximum and minimum, the pattern in Murcutt's inequality genotype becomes more visible (fig. 17).

In the next stage of the analytical process a "statistical archetype" is constructed. However, in the present context a set of five houses, with a total of fifty defined nodes, is not sufficiently large to derive a meaningful mathematical trend. While the past precedents for this method have had, on occasion, data sets of similar size, the more compelling results using this method tend to come from sets with at least double the number of nodes. While acknowledging this weakness, the present paper uses broad trends to construct a visual analysis diagram. This diagram is still called, for methodological consistency, a "statistical archetype" but the reader should remember that its construction is less robust than would otherwise be desired.

Within the visual analysis stage of this paper a range of observations recorded the relative frequency, in a set of five designs, where certain permeability and hierarchical patterns were identified. This information is used to construct a table of tendencies for structural and programmatic properties to occur in a project. The percentage probabilities were determined directly from the designs, thus if four of the five designs featured a particular pattern, then that is described as an 80% chance. The patterns identified through visual analysis in this paper are as follows:

1. There is a 80% chance that there will be an entry "ring" configuration founded on the exterior carrier, which encompasses, in order of probability, a hallway, a veranda and a dining room, or dining, kitchen and living area combination;

2. There is an 80% chance that the hallway will be the starting point for an arborescent branching structure leading to two bedrooms and a bathroom;

3. There is a 60% chance that there will be a mezzanine structure and a 40% chance that it will be for the third bedroom and will be accessed from a ground floor open plan dining/living area;

4. Given the above observations, the depth of the JPG, with exterior as carrier, is likely to be three layers (0-3);

5. While larger plans are a minority condition (40%), if they are required, a second hall will be added as the starting point for an additional branching structure, which may encompass either a garage wing, or a guest bedroom wing.

Taking all of these rules into account, it is possible to construct a partial "statistical archetype", or more correctly a trend diagram, for an early Murcutt house (fig. 18). Note, in this graph, the recurrent theme of the entry ring structure for the public parts of the house, which includes a hall, the starting point for a secondary arborescent structure that governs access to the more private zones in the house.

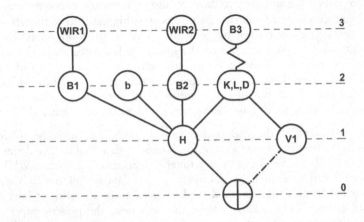

Fig. 18. JPG Statistical Archetype, with exterior as carrier, for an early Murcutt house

It is possible to interpret the Statistical Archetype graph visually from two divergent points of view, those of the stranger, and of the inhabitant [Hillier and Hanson 1984]. For example, the stranger entering this house has, by virtue of the entry ring, immediate access to the entire domestic core. In contrast, the inhabitant has their privacy (bedrooms, bathrooms, walk-in-robes) strongly protected through certain control zones (the origin points of the bush-like structure), but once within the body of the house they may take advantage of its open structure. This might seem to be an inversion of the anticipated social structure of a house, that is, a configuration that provides limited entry and access for the stranger and a more open interior for the inhabitant. But before accepting this conclusion, it must be remembered that the five houses are all on isolated rural properties where strangers are extremely rare. Indeed, it is more common for these houses to have guest spaces for people who have been invited to stay with the owners for a few nights. This explains both the strongly hierarchal nature of the private spaces (separating owners from guests) and the relative lack of concern about the programmatic depth afforded to strangers.

Leaving aside the comparative visual analysis, the five houses may be compared mathematically, through a review and comparison of their relative difference factors, $H^*$ which provides a measure of the degree of differentiation between spaces in terms of integration. It is also useful for comparative purposes because it normalises results. Once the $H^*$ figure is determined, then it is interpreted as follows:

> The closer to 0 the difference factor, the more differentiated and structured the spaces …; the closer to 1, the more homogenised the spaces …, to a point where all have equal integration values and hence no configurational differences exist between them [Hanson 1998: 30-31].

The $H^*$ results for the five houses are as follows: the Marie Short House $H^*$=0.56; Nicholas House $H^*$=0.61; Carruthers House $H^*$=0.703; Fredericks Farmhouse $H^*$=0.799; and Ball-Eastaway house $H^*$=0.382. Only the Ball-Eastaway house has a spatial configuration ($H^*$ value) that falls into the category of "differentiated", "deliberative" or "strong" genotypes, but even that result is much closer to the middle range. The remaining four houses could more accurately be described as "homogenised", "loose" of "weak" genotypes with little structural differentiation between the spaces (which suggests that they could equally be bedrooms, bathrooms, kitchens or garages).

## Conclusion

It was Philip Drew who noted that while the Marie Short, Nicholas, Carruthers and Fredericks houses share a number of external formal similarities, they are also quite different "in the arrangement of their spaces" [1985: 92]. Equally diplomatically, Beck and Cooper observe that the "Murcutt hallmark, the long plan" is seemingly able to be adapted "to any given programme" [2002:11]. These are rare instances where a critic implies that perhaps there is a lack of connection between the rigorous formal or geometric strategy that dominates the exterior and the somewhat loose planning strategy, or topology, of the interior. Farrelly is slightly more forthright than most when she argues that while Murcutt's "forms are universal [and] rationalist", the configuration of his plans "is particular, empirical and contingent" [1993: 21]. The JPG analysis in the present paper uncovered many examples of what Farrelly describes as "particular" or "contingent" spatial planning. Indeed, hallways dominate many of the plans and several secondary spaces, including garages and bedrooms, are also more critical to the plan and its circulation than the living or dining spaces. Similarly, almost all of Murcutt's loft spaces are accessed from unusual places (living rooms, utility hallways), a factor which consistently generated unlikely permeability results.

Murcutt himself notes that a simple form does not necessarily imply the presence of a simple interior. "The house [may be] very simple. But remember simplicity is the other face of complexity" [2007: 26]. In this statement Murcutt suggests that the apparent simplicity of the exterior form of a building may mask a more complex interior. This is certainly the case with the interiors of the five houses investigated in this paper. With the possible exception of the Ball-Eastaway, "train carriage" hierarchical plan, all of the rest of the spatial configurations were both more complex and less predictable than the canonical literature suggests. Certainly Pallasmaa's claim, cited early in this paper, that the form and spatiality of the houses are perfect reflections of each other, is impossible to maintain in light of the present research. Murcutt's spatial planning, while generally neatly zoned into served and servant spaces, is clearly not the primary or even the secondary driver of his design approach.

From a Space Syntax perspective this result echoes Bafna's observations about the domestic architecture of Mies van der Rohe, an architect Murcutt has often been favourably compared with. In both cases, rather than the geometry of the building being subservient to the internal genotype, geometry is the starting point for formulating the limits and constraints of a design, within which a program is forced to fit. As Bafna notes, one of the key anomalies in the early use of the JPG as that it assumed that "spatial organization has generally been seen as happening decisively within an entirely topological space, with geometry providing an opportunity for embellishments" [2001: 20.15]. Instead, Bafna suggests, it may be better to imagine that design progresses from a geometrical starting point which "permits, and indeed makes possible, a great deal of topological variation within certain restrictions" [2001: 20.15]. Thus, in the architecture of Mies van der Rohe and Glenn Murcutt, form does not follow programmatic function at least. Indeed, as many analysts have noted, the more "functional" a space, the less capacity it has to adapt to changing social and cultural conditions [Blake 1974; Brolin 1976].

A good example of this can be seen in the way in which Murcutt uses narrow slivers of space to connect parallel pavilion forms (something that occurs in three of the houses considered in this paper). In two of the cases (the Marie Short House and the Fredericks House) the space becomes a corridor, whereas in the third (the Nicolas House) it is largely merged with the rest of the open plan. Similarly, the steeply pitched roofs of these houses conceal mezzanine rooms on some occasions but not on others. In each of these examples – the connecting sliver and the inhabited roof – a review of the external form alone cannot be used to predict the relative depth, or social structure, of the interior.

Ultimately, because of the similarities between the architecture of Murcutt and Mies, it is not surprising that the present paper has reinforced the findings of Bafna [1999]. Both architects' minimalist aesthetic compositions clearly require some compromise. What would be more interesting for future researchers to test would be the inequality genotypes produced for houses designed by architects who have openly expressed a primary concern with program; Charles Moore, Christopher Alexander and Patkau Architects all fit into this category. Finally, an alternative research direction would be to expand the present paper to include new, larger and more complex rural houses by Murcutt. The larger programs of these houses provide further opportunities for considering the relationship between form and planning, or geometry and topology.

### Acknowledgments

An ARC Fellowship (FT0991309) and an ARC Discovery Grant (DP1094154) supported the research undertaken in this paper. All of the images in the paper are by Romi McPherson for the author.

### References

ALEXANDER, Christopher. 1966. A City is Not a Tree; Part I and Part II. *Design* **206** (February 1966): 46-55.

ALEXANDER, Christopher, S. ISHIKAWA and M. SILVERSTEIN. 1997. *A Pattern Language: Towns Buildings, Construction.* Oxford: Oxford University Press,

BAFNA, Sonit. 1999. The Morphology Of Early Modernist Residential Plans: Geometry and genotypical trends in Mies van der Rohe's designs. *Proceedings of the Second International Symposium on Space Syntax,* vol. 1, 01.1-01.12. Available online a http://217.155.65.93:81/symposia/SSS2/sss2_proceedings.htm. Last accessed 15 June 2011.

———. 2001. Geometric Intuitions of Genotypes. *Proceedings of the Third International Symposium on Space Syntax,* 20.1-20.16.

———. 2003. Space syntax: a brief introduction to its logic and analytical techniques *Environment and Behavior* **35**, 1: 17-29.

BAKER, Geoffrey. 1996. *Design Strategies in Architecture: An Approach to the Analysis of Form* New York: Van Nostrand Reinhold.

BECK, Haig and Jackie COOPER. 2002. *A Singular Architectural Practice.* Melbourne: The Image Publishing Group.

BLAKE, Peter. 1974. *Form Follows Fiasco: Why Modern Architecture Hasn't Worked.* Boston Little Brown and Co.

BROLIN, Brent C. 1976. *The Failure of Modern Architecture.* London: Studio Vista.

CHING, Francis D. K. 2007. *Architecture: Form, Space and, Order.* Hoboken, NJ: John Wiley and Sons.

DOVEY, Kim 2010. *Becoming Places: Urbanism / Architecture / Identity / Power.* London: Routledge.

———. 1999. *Framing Places: Mediating Power in Built Form.* London: Routledge.

DREW, Philip. 1985. *Leaves of Iron: Glenn Murcutt.* Sydney: The Law Book Company.

———. 1986. House with a Roof Like a Butterfly Spreading Its Wings. *Architecture* **75** (September 1986): 58-60.

———. 2001. *Touch this Earth Lightly.* Sydney: Duffy and Snellgrove.

FARRELLY, Elizabeth M. 1993. *Three Houses.* London: Phaidon Press.

FRAMPTON, Kenneth. 1995. *Studies in Tectonic Culture: The Poetics of Construction in Nineteenth And Twentieth Century Architecture.* Cambridge, MA: MIT Press.

———. 2002. The Architecture of Glenn Marcus Murcutt. http://www.pritzkerprize.com/laureates/2002/essay.html. Last accessed 15 June 2011.

———. 2006. *Glenn Murcutt, Architect.* Sydney: 01 Editions.

FROMONOT, Francoise. 1995. *Glenn Murcutt Buildings and Projects.* London: Thames and Hudson.

GELERNTER, Mark. 1995. *Sources of Architectural Form: A Critical History of Western Design Theory.* New York: St. Martin's Press.

GUSHEH, Maryam, Tom HENEGHAN, Catherine LASSEN and Shoko SEYAMA. 2008. *The Architecture of Glenn Murcutt.* Tokyo: TOTO.

HANSON, Julienne. 1998. *Decoding Homes and Houses.* Cambridge: Cambridge University Press.

HANSON, N. L. R. and Tony RADFORD. 1986a. Living on the edge: a grammar for some country houses by Glenn Murcutt. *Architecture Australia* **75**, 5: 66-73.

———. 1986b. On Modelling the Work of the Architect Glenn Murcutt. *Design Computing* (1986): 189-203.

HILLIER, Bill and Julienne Hanson. 1984. *The Social Logic of Space.* New York: Cambridge University Press.

HILLIER, Bill. 1995. *Space is the Machine.* Cambridge: Cambridge University Press.

HILLIER, Bill and Kali TZORTZI. 2006. Space Syntax: The Language of Museum Space. Pp. 282-301in *A Companion to Museum Studies,* Sharon Macdonald, ed. London: Blackwell.

MACMAHON, Bill. 2001. *The Architecture of East Australia.* London: Edition Axel Menges.

MAJOR, Mark David and Nicholas SARRIS. 1999. Cloak and Dagger Theory: Manifestations of the mundane in the space of eight Peter Eisenman houses. *Proceedings of the Second International Symposium on Space Syntax,* vol. 1, 20.1-20.14. Available online at http://217.155.65.93:81/symposia/SSS2/sss2_proceedings.htm. Last accessed 15 June 2011.

MARKUS, Tom. 1987. Buildings as Classifying Devices. *Environment and Planning D: Society and Space* **14**: 467-484.

———. 1993. *Buildings and Power.* London: Routledge.

MURCUTT, Glenn. 2007. Thoughts on the Ecology of Architecture. *Architecture and Urbanism* **8**(443): 26.

OSMAN, Khadiga M. and Mamoun SULIMAN. 1994. The Space Syntax Methodology: Fits and Misfits. *Architecture & Behaviour* **10**, 2: 189-204.

OSTWALD, Michael. J. 1997. Structuring Virtual Urban Space: Arborescent Schemas. Pp. 451-482 in Intelligent *Environments – Spatial Aspects of the Information Revolution,* Peter Droege, ed. Amsterdam: Elsevier.

———. 2011. The Mathematics of Spatial Configuration: Revisiting, Revising and Critiquing Justified Plan Graph Theory. *NNJ* **13**, 2 (Summer 2011): 445-470.

PALLASMAA, Juhani. 2006. The Poetry of Reason. Pp. 15-20 in Kenneth Frampton, *Glenn Murcutt, Architect.* Sydney: 01 Editions.

PEPONIS, John, Jean WINEMAN, Mahbub RASHID, S. HONG KIM and Sonit BAFNA. 1997. On the description of shape and spatial configuration inside buildings: convex partitions and their local properties. *Environment and Planning B: Planning and Design* **24**: 761-781.

SPENCE, Rory. 1986. At Bingie Point. *The Architectural Review* **1068** (February 1986): 40-43.

UNWIN, Simon. 2003. *Analysing Architecture.* London: Routledge.

ZAKO, Reem. 2006. The power of the veil: Gender inequality in the domestic setting of traditional courtyard houses. Pp. 65-75 in *Courtyard Housing: Past, Present and Future*, Brian Edwards, Magdo Sibley, Mohamad Hakmi and Peter Land, eds. New York: Taylor and Francis.

## About the author

Professor Michael J. Ostwald is Dean of Architecture at the University of Newcastle (Australia), Visiting Professor at RMIT University (Melbourne) and a Professorial Fellow at Victoria University Wellington (New Zealand). He has a Ph.D. in architectural history and theory and a higher doctorate (D. Sc) in the mathematics of design. He has lectured in Asia, Europe and North America and has written and published extensively on the relationship between architecture, philosophy and geometry. He has a particular interest in fractal, topographic and computational geometry and has been awarded many international research grants in this field. Michael Ostwald is a member of the editorial boards of the *Nexus Network Journal* and *Architectural Theory Review* and he is co-editor of the journal *Architectural Design Research*. He has authored more than 250 scholarly publications, including 20 books. His recent books include *The Architecture of the New Baroque* (2006), *Residue: Architecture as a Condition of Loss* (2007), *Homo Faber 1: Modelling Design* (2008), *Homo Faber 2: Modelling Ideas* (2008) and *Understanding Architectural Education in Australasia* (2009). He is also co-editor of *Museum, Gallery and Cultural Architecture* in Australia, New Zealand and the Pacific Region (2007).

# Aineias Oikonomou

25, Iasonos Street
Vouliagmeni, 166 71
Athens GREECE
aineias4@yahoo.com

Keywords: architect's cubit,
ottoman architecture, design
analysis, northern Greece,
structural systems, symbolism,
diagonals, geometry, geometric
systems, measuring systems,
module, morphology, number,
proportion, proportional
systems, symmetry

## Research

# The Use of the Module, Metric Models and Triangular Tracing in the Traditional Architecture of Northern Greece

**Abstract.** This paper presents the analysis of the design and construction of the traditional houses in northern Greece and the correlation with the use of the metric system of the architect's cubit and the existence of architectural tracing and proportions. Of particular significance is the detection of certain metric models widely applied in the urban centres of northern Greece during the nineteenth century, and the general application of the Pythagorean theorem on the triangular tracing. The analysis was based on the application of the functional and constructional module on plans and façades of houses in order to investigate its influence on the standardisation of the design of the spaces and their elements, as well as of the construction. This makes it possible to reconstruct the design principles that characterise the traditional houses in the Ottoman Empire and the proportions of their parts, as well as to identify similarities and differences between the various urban centres of northern Greece. Another finding of the research involves the way in which the structural grid is applied to the design of the façades and to the placement of the structural elements and the openings of the timber-frame construction. As a result, conclusions are drawn concerning the role and the effect of the measuring system and the various metric models on the constructional rationale and on the development of the form of the traditional house.

## 1 Previous studies of tracing and proportion

The tracing and proportions that define the structure and the form of buildings from early on until the present day, have been the object of many previous studies. For the architecture of the Ottoman period, and specifically in the area of the Balkan Peninsula, the studies that are related to this paper are summarised below.

N.K. Moutsopoulos analysed the existence of geometrical harmonic tracing in the façades and the interior spaces of large mansions in Kastoria with the use of circles and diagonals. The façades are subdivided into smaller zones and parts, from which smaller sizes and proportions are derived (see [Moutsopoulos 1988: 126-128]). In his earlier studies, Moutsopoulos [1956] applied geometrical tracing in sections of Byzantine inscribed cross-shaped churches and showed that there exist constructional tracings that define the proportions of their parts (see [Moutsopoulos 2010: 100-136] and [Konstantinidis 1961: 51-52]). Similar studies concerning medieval churches in Bulgaria were presented by I. Popov [1955] and G. Kozukharov [1974]. Popov investigated the proportions using diagonal tracing and squares (quadratura or ad quadratum), while Kozuharov used equilateral and right-angled triangles in order to explore circular and arched tracing.

For Bulgaria, R. Anguelova [1993] and M. Cerasi [1988] cite the studies of P. Berbenliev, who investigated harmonic tracing following the principles of the golden section in the façades of eighteenth- and nineteenth-century houses. It is also mentioned that "the masters of Bratzigovo ... traced their plans directly on the site, which would sometimes result in harmonic relation and even in golden rule proportions" [Cerasi 1988: 93].[1] Furthermore, Cerasi mentions that "the analysis of plans and elevations of buildings by Balkan masters has shown the existence of proportions and tracing inscribed in circles",[2] but stresses that "these are called up by the fundamentally empirical modular measurement systems as well as by the general rationality of Ottoman architectural practice" [Cerasi 1988: 95]. Moreover, concerning constructional tracing and proportions, Stoichkov draws the plan of a nineteenth-century mansion in Koprivchititsa, Bulgaria, overlaying a modular grid of 90 cm along with axes of symmetry and tracing [Stoichkov 1977: 58]. Cerasi redraws the above-mentioned plan on an *arşin* grid (75.8 cm), which is based on the actual module used by master-masons during that period [Cerasi 1988: 99]. All in all, Cerasi analyses the metric system of the period of the Turkish domination and its use by the builder guilds. He also presents plans of public baths and private houses of the seventeenth and eighteenth centuries from Topkapi, which are laid out on a grid, and thus proves that the basic module in use is the architect's cubit (*mimar arşin*) [Cerasi 1988: 93]. Similar plans of public baths and religious buildings drawn in scale on an arşin grid are also presented by G. Necipoğlu-Kafadar [1986].

Apart from the above, B. Sisa analysed plans and sections of Protestant wooden bell towers in the Carpathian basin using as a starting point the medieval theory of the constructional principles of the circle-square-triangle,[3] which provides evidence that carpenters did not use drawings but applied these principles directly during the construction stage [Sisa 1990: 327]. Sisa's drawing analysis shows that these timber structures are based on a constructional grid with specified proportions [Sisa 1990: 338, 347].

Based on this bibliographical review, it can be seen that apart from the studies by Stoichkov and Cerasi, who apply a uniform constructional grid to house plans, and Necipoğlu-Kafadar, who presents original drawings, the rest of the studies do not use or present a given module but apply tracing of circles and diagonals that subdivide the façades or sections into smaller parts, without mentioning the use of the module.

In relation to the above, the contribution of the present study lies mainly in the investigation of metric proportions and the use of the module in the design of traditional houses and the identification of specific metric models and building types in combination with architectural tracing, which are largely applied in the urban centres of Northern Greece during the eighteenth and nineteenth centuries. It is shown that constructional tracings exist, but only concerning the plan and its tracing on the ground whereas the proportions of façades and heights derive from the consequent subdivision and not from diagonal tracing.

## 2 Methodology

The starting point of this study is the author's Ph.D. thesis [Oikonomou 2007] which involved the analysis of the typological, morphological, constructional and environmental characteristics of forty traditional houses in Florina, in northwestern Greece. An important part of this work was dedicated to the investigation of the application of the module in houses of the nineteenth century.[4] The research led to scientific paper presented in Greece [Oikonomou and Dimitsantou-Kremezi 2009]. More recent work [Oikonomou et al. 2009] extended the research in order to include eight houses from different towns and settlements in north-western Greece. Four of them were situated in Florina and were included because of their unique characteristics, whereas the rest of the examples were from Ano Kranionas in the Korestia area, Trikala, Oxia on Mount Vitsi, and Kastoria. The selection was based on the fact that their plan design is derived from the same building type design, that of the house with the central (cross-shaped) *sofa*.

The current work is a further extension of this approach to selected traditional houses in the wider area of Northern Greece. This investigation is based on the application of the functional and constructional module (a modular grid) on original or redrawn plans and façades of selected typical houses. To compile the present study, a total of one hundred houses from different towns and settlements in Northern Greece were selected and analysed, twenty of which are presented here. Most of them are situated in Florina, Kastoria, Siatista, Eratyra, Kozani, Veria, Serres, Xanthi, Ioannina, Arta, Trikala and Ampelakia. Apart from these, two remaining Ottoman monuments of Florina are also included in the study. The aim of this analysis is to investigate the influence of the module on the standardisation of the design of the spaces and their elements, as well as on the construction.

Before beginning, it will be helpful for the reader to define the main elements that characterise the houses here. These are *oda*, room; *sofa*, central hall; *eyvan*, covered patio or alcove; *hayat*, a living space.

## 3. Investigation of the application of the module in the traditional architecture of northern Greece

### 3.1 The metric system of the *mimar arşin*, or architect's cubit

The measuring of plots and the construction of buildings in the Ottoman Empire was based on the *mimar arşin*, or architect's cubit. The arşin is divided into 24 *parmak*, or fingers; the finger is divided into 12 *hatt*, or lines; and the line is divided into 12 *nokta*, or points [Cerasi, 1988: 92]. The arşin is also divided into two equal parts called *kadem* [Özdural, 1988: 113].

The Ottomans used the arşin from the beginning of the sixteenth century, though its precise measurement varied. Alpay Özdural mentions that the imperial measuring unit was equal to 72.1 cm in 1520; 73.4 cm during the last quarter of the sixteenth century [Özdural 1988: 103]; and 76.4 cm during the third quarter of the eighteenth century. The arşin of 75.8 cm was the fourth and final version, assigned as the imperial measuring unit by Selim III, during the years 1794-95, and remained in use until 1934, when it was replaced by the metre [Özdural 1988: 106].

The arşin, similar to the old English yard, was both an instrument and a measuring unit [Cerasi 1988: 92]. As an instrument, it was usually made of ebony, wood, bronze, copper or bone and could be folded into two parts (kadem) or four parts. As mentioned, it was subdivided into parts of 12 or 24 [Özdural 1988: 113].

### 3.2 The module and the grid

From earlier studies [Oikonomou 2007, 2009], it was seen that the oda usually have a square plan with dimensions 5x5 or 6x6 arşin. The sofa has a rectangular plan with its small dimension (width) equal to 3, 4, or 5 arşin. When the hall results from symmetrical repetition, its width is equal to 7 or 8 arşin (2×3.5 or 2×4). The sizes of the oda and sofa are usually combined, resulting in many different possible variations. It should be noted that in houses with large odas with dimensions 6×6 arşin, the sofa's width is never smaller than 4 arşin. The same applies to four-room houses with an inner sofa, even when the odas are 5×5 arşin. In general, from the research presented in this paper, it can be seen that there is a tendency to use whole multiples of the arşin.

Concerning the placement of windows, it should be noted that the module is equal to the width of the openings. As a result, in an oda with dimensions 5×5 arşin, the windows are placed either at intervals of 1 arşin or at intervals of 1 kadem. In the first case, two windows are placed on each façade, whereas in the second one, each façade has three windows (fig. 1).

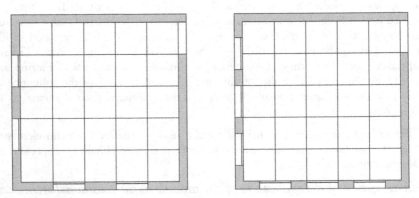

Fig. 1. The placement of windows according to the constructional and functional module

### 3.3 The basic metric models

The metric models, which can be found in the architecture of northern Greece, as well as in the urban centres of the wider Balkan area, are possibly based on plans that existed in Istanbul.[5] These plans were drawn on rid paper with a module based on the architect's cubit and at a scale of 1:48 [Cerasi 1988: 93]. The module is related to the dimensions of the spaces and to the spacing of the structural elements. In the earlier designs, the walls are depicted in great detail, while in the more recent ones, the walls are depicted with one single line and the decorative elements in great detail [Cerasi 1988: 93, pictures 1 and 2].

In the architecture of northern Greece, the two most common metric models that are observed are:

a) The house with odas of 5×5 arşin and sofa with a width of 3 arşin (figs. 2, 14);
b) The house with odas of 6×6 arşin and sofa with a width of 5 arşin (figs. 9, 20-21)

The first model is usually used in houses with two odas and a sofa, whereas the second one in houses with four odas and an inner sofa.

The following models are also commonly found:

c) Houses with odas of 5×5 arşin and sofa with a width of 4 arşin (figs. 5, 15);
d) Houses with odas of 6×6 arşin and sofa with a width of 4 arşin (fig. 16).

The house model a-5/3/5 can be adjusted on the plot and the needs of the owner. The subtraction of an oda results in a non-symmetrical house with one oda of 5x5 arşin and a sofa of 3x5 arşin (model a1-5/3) with a total length of 9 arşin. Respectively, the doubling of the sofa, leads to a house with two odas of 5x5 arşin and a twin sofa of 3x5 arşin (model a2-5/3/3/5). The total length in this case is 17 or 18 arşin, depending on the thickness of the external walls [Oikonomou and Dimitsantou-Kremezi 2009: 311, fig. 2].

In the house model c-5/4/5, the doubling of the sofa can lead to a house with four 5x5 arşin odas and a twin sofa of 4x11 arşin (model c1-5/4/4/5). The total length in this case is 19 arşin and the width 12 arşin [Oikonomou et al. 2009: 555, fig. 13-14].

At this point, a standardisation concerning the total length of the building façades should be noted. Usually, the length of the house is 14 arşin (model a-5/3/5 and c-5/4/5) or 17 arşin (model b-6/5/6). Furthermore, there are many houses with a length of 16 arşin (model c-5/4/5 and d-6/4/6). Finally, there are large mansions with a total length of 21 or 23 arşin (model f-6/8/6 or model g-7/7/7).

This standardisation is closely related to the standard thickness of the walls. For instance, a house with two odas of 5x5 arşin, a sofa 3 arşin wide and light-weight timber-frame walls of 19 cm (1/4 a.) has a total length of 14 arşin (5+3+5+1), whereas a house with two odas of 5x5 arşin, a 7 arşin wide hayat, thick exterior walls (57cm – 3/4 a.) and light-weight interior walls has a total length of 19 arşin (5+7+5+2). In conclusion, the choice between thick or thin walls defines the final dimensions (length and width) of the house (fig. 2).

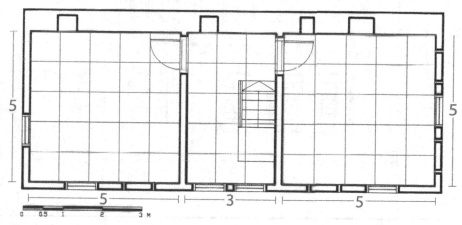

Fig. 2. Plan of a building (Mavroudis house) with rooms of 5x5 arşin and a sofa of 3x5 arşin

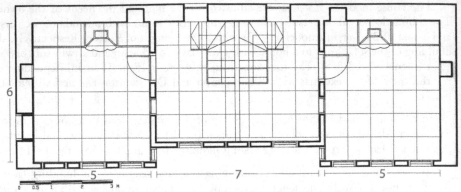

Fig. 3. Plan of a building (Anagnostopoulou house) with rooms of 5 x 5 arşin
and a hayat of 7 x 5 arşin

Fig. 4. Façade of Ioannou Mansion in Florina.
The upper floor height is 4 arşin and the ground floor height is 3.5 arşin

Fig. 5. Façade of Chatzitzotzas Mansion in Florina.
Redrawn from plans by A. Stoios and M. Papagiannakis

Finally, it should be noted that the metric system of the architect's cubit is also applied to the building façades, where, similar to the building plans, a standardisation of sizes is also observed. The total height of the houses, without the roof, is usually 7 arşin. Nevertheless, there are many houses with a height of 8 or 9 arşin, which have a raised ground floor. The upper storey height is 3.5 or 4 arşin, while the ground floor height is 3 or 3.5 arşin. These dimensions correspond to the exterior dimensions of the spaces, the dimensions of the structural elements and, consequently, to the dimensions of the façades. Finally, the upper storey windows have a sill height of 1 arşin and a clear height of 2 arşin, whereas in the ground floor, they have the same sill height, but a total height of 2 arşin including the cornice. In this way, the placement of the openings on the façades is characterised by a relationship of a-2a-a, where a is equal to the window's width and the sill height, and 2a to its height (figs. 4-5).

### 3.4 The use of triangular tracing

The metric models that are analysed contribute to the standardisation of the construction of the traditional house. Most of the buildings presented have a fully rectangular plan. At the same time, the constructional and functional module is applied accurately, not only in the placement of the basic structural elements and the openings, but also in the dimensions of the spaces. All the above lead to the conclusion that before the construction of the house, certain basic architectural tracing, which assumes basic knowledge of geometry, took place.

For the tracing of the rectangular perimeter of the house and its basic walls, the Pythagorean theorem with the basic right-angle 3-4-5, 5-12-13 and 8-15-17 triangles[6] is applied with the use of *knotted ropes*. Stoichkov has proven that master builders in Bulgaria used the 3-4-5 and 5-12-13 triangles during the nineteenth century [Stoichkov 1977: 126, 145]. The same methods certainly apply in Northern Greece during this period.

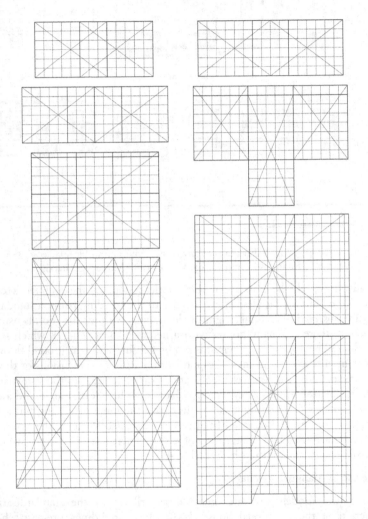

Fig. 6. Use of the module and triangles 3-4-5, 6-8-10, 9-12-15 and 5-12-13 in the tracing of the metric models

The 3-4-5 triangle can be multiplied by various factors (2, 3, 3.5, 4, 5, 5.5, 6, etc.) t produce triangles of the necessary dimension; thus multiplication by a factor of produce the 6-8-10triangle; by a factor of 3, 9-12-15; and so forth. The 5-12-13 triang can be divided by a factor of 2 to produce a triangle 2.5-6-6.5, whereas the 8-15-1 triangle can also be divided by the same factor to produce a triangle 4-7.5-8.

Furthermore, the use of circular auxiliary tracing can not be ruled out, especially for houses with odas of 5 arşin and a sofa of 3 arşin. The numbers 5-3-5 are related to the subdivision of a line with a length of 13 into two parts with lengths of 8 and 5 (extreme and mean ratio) and the Fibonacci series (see [Konstantinidis 1961: 190-191]).

The following diagrams include observations concerning the various metric models and their way of tracing (fig. 6). In this way, the relationship with the above-mentioned characteristic right-angled triangles is investigated and conclusions concerning the possible tracing of the spaces of the houses and their design can be drawn. Finally, the possible influence of the characteristic 3-4-5 triangle in the placement of the diagonal structural parts of the timber frame and in the form of the small pediment on the main façade is also investigated. Noting that tan 53° is approximately equal to 4/3 and that tan 37° is approximately equal to 3/4, the fact that the diagonal elements are in many cases placed at an angle of 53° and that the pediment often has a slope of 37° suggest that this triangle might have also been used in the construction of the façades, independent of the tracing of the plan (fig. 7).

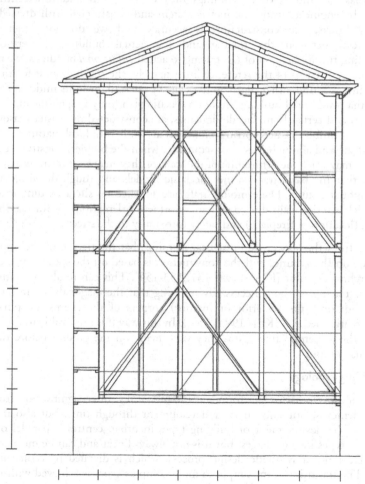

Fig. 7. Use of the module and basic triangle 3-4-5 in the construction of the timber-frame structure

At this point it is useful to site the definition given by Jouven concerning the (harmonic) tracing:

> A harmonic trace is a geometric figure whose design coincides with the main lines of a building (or what amounts to the same, its graphic representation in geometrical projection). The choice of this geometric figure, conceived at the same time as the building by the architect, is based on the remarkable properties that are included in it [Jouven 1951: 7].

Jouven also explains:

> We mean by main lines those whose knowledge is sufficient to determine the elements of construction (perimeter, axes, lines of change, etc.). We can say in a quick way that they are the main lines of a first rough drawing of the building [Jouven 1951: 7].

In well-known *academic* plans of important buildings of the Renaissance period or the neoclassical and eclectic buildings of the eighteenth and nineteenth centuries, designed by engineers using the metric system and constructed with dressed stone and fired-clay bricks, the construction is accurate and we do not generally observe discrepancies between the tracing and the actual building. Nevertheless, when investigating the application of the tracing to actual *traditional* buildings, we must expect discrepancies in terms of the reference lines; namely, the starting point for the tracing of the building is found either on the outer edge of the wall, or in its middle. This is due to the fact that traditional buildings have no specific plan (only rough drawings) (see [Cerasi 1988: 93]) and certainly no façade drawings, are constructed by master masons based on principles of timber-framed construction, and are built with local, natural materials, such as field stone and adobe bricks. Consequently, when the builders construct a thick adobe wall, they may trace the centre line of the wall, or they may start from its exterior corner. It seems that the basic tracing and the basic models and rough drawings involve only light-weight structure. These model drafts are adjusted on site according to the need to create thick exterior walls due to climatic (northern orientation), or for reasons related to function (location of fireplace) and safety (proximity to the street).

Apart from the above, we must bear in mind that the tracing of the upper stories is also made on the ground, and the construction is based on the specific grid and module of the architect's cubit [Kizis 1994: 131, 153-154]. This can result in a rather complex triangular tracing, without necessarily meaning that the diagonals are in fact traced on the ground: only the (exterior or interior) corners of the rooms are pin-pointed by wooden marks (see also [Kizis 1994: 154]). In this way it is impossible to find any sign or mark of the presented tracings as they were made on the ground before the buildings were actually built.

## 3.5 Building typology

The house types that are found in urban centres differ significantly from those in rural settlements, not only in their development through time, but also in their basic structure. The development of building types in urban centres is parallel or later than that of the types in the villages, but it is not always linear and can be more intricate. We could talk about a synthetic design process, which is dictated by economic or spatial factors. The transition from simple to more complex types is achieved with the addition (multiplication) of spaces during the eighteenth and early nineteenth centuries (see [Kuban 1995: 105]). Nevertheless, during the second half of the nineteenth century there are house plans that derive from the reduction of the number of spaces – namely the subtraction of one oda – from a closed, complete type, such as that of the house with the central sofa.

The master-builder proposes the application of the building type that is appropriate not only for the owner's economic situation and needs, but also to the size of the plot, which, in the urban centres, usually has a limited street front. These parameters lead to the adjustment of the basic model type with the subtraction of living spaces, and as a result variant types, which can be considered, for example, as the two thirds (2/3) of a complete and established type, occur. In other instances, the repetition of an element, such as the sofa, leads to larger, twin-type houses (fig. 8).

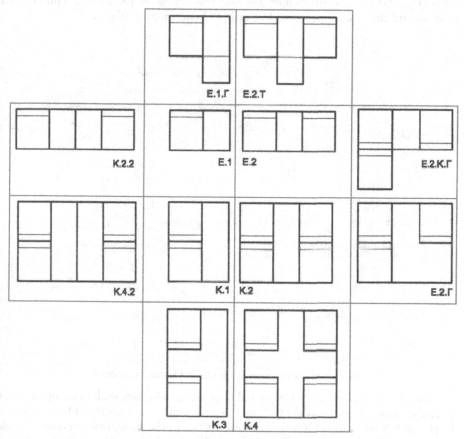

Fig. 8. Typological considerations for the traditional houses of the nineteenth century

## 3.6 Building construction

The light-weight walls of the upper storey (*çatma*) are thin (19 cm) and formed by a complex timber frame, which is filled with adobe bricks and, in some cases, small stones or fired-clay bricks. The basic, vertical structural elements (*direk*) are positioned in the corners of the spaces and between the openings, at regular intervals of about 150 cm (2 arşin). In between, horizontal elements are placed at heights of 75 and 225 cm (1 and 3 arşin). This leads to the creation of large and small frames. In the former, openings with a width of 75 cm (1 arşin) are placed at regular intervals relative to the dimensions of the interior space. In the latter, smaller vertical elements are placed over and under the window. Finally, diagonal elements of triangulation are placed in all the frames in order to increase the rigidity of the structure.

The constructional grid of the main façade is directly related to the plan and is very interesting. The basic structural elements are always placed at regular intervals, while the windows are placed symmetrically with respect to the interior space. In this way, in the upper storey of a typical house with odas with an interior dimension of 5 arşin (379 cm) and an exterior of 5.5 arşin (417 cm), the four basic direk are placed at 1.75 arşin (133 cm) intervals, whereas the three openings are placed at 0.5 arşin (38 cm) intervals. Similarly, in a typical house with upper-storey odas of 6 arşin (455 cm) interior and 6.5 arşin (492 cm) exterior dimensions, the four basic direk are placed at 2 arşin (152 cm) intervals and the three openings at 0.75 arşin (57 cm) intervals (fig. 9).

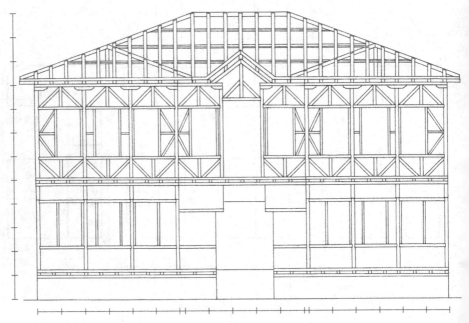

Fig. 9. Typical construction of a house in Florina, Northern Greece

The structural elements of the ground floor are usually thick walls made of local ston or adobe bricks. These walls are 0.75 to 1 arşin (57 to 75 cm) thick and have an averag height of 3.5 to 4 arşin (265 to 300 cm) and comprise horizontal structural wooden elements (ties) at intervals of 1 arşin (75 cm). This construction is typical in most town and rural settlements. Finally, it should be noted that in Florina, where a local structura system exists, the ground floor walls of increased thickness include direk at regula intervals (4, 5 or 6 arşin) that follow the structural module and correspond to the interic partition walls. On the upper storey, the direk are placed in the corners and the middle c every space (at 2, 2.5 or 3 arşin intervals). The main reason for the development of th system is the fact that the construction material (local river stone) is relatively weak an the subsoil is unstable because of the river bed.

# 4 Presentation of the study

## 4.1 The Ottoman monuments of Florina

The Ottoman bath in the town of Florina has three easily discernable building phases, spanning a period of three centuries. The core of the bath can be dated according to typology (bath with small dimensions, see also [Kanetaki 2004]) and the measuring system of its construction (9 by 6.75 arşin, with an arşin equal to 73.4 cm, mentioned by Ibn Ma'ruf for the period 1569-74), to the end of the sixteenth or the beginning of the seventeenth century [Özdural 1988]. This is in agreement with the mention of the bath by Evliya Çelembi (1661) [Dimitriadis 1973] with the interior morphological elements (arches, squinches and dome), and the careful construction of the external stonework (dressed stones and cornerstones). The tracing of this core is based on the use of a triangle 3-4-5 multiplied by a factor of 2¼ to produce a triangle measuring 6¾ - 9 - 11¼ arşin (fig. 10).

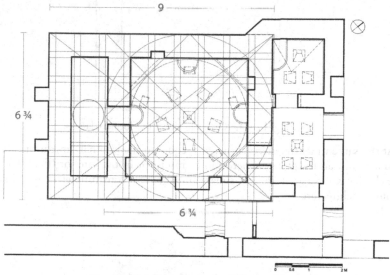

Fig. 10. The Ottoman bath of Florina. Plan

At the same time, the octagonal tracing of the central space and squinches is based on the use of a circle with a diameter of 6¾ arşin, and in this way the interior diagonal is equal to the exterior dimension. The transversal timber frame-work (*klapes*) are also placed at approximately 1 arşin (73.4 cm) intervals, whereas the fired-clay bricks used in the interior side of the walls and the dome are based on 1/3 arşin (dimensions of 23 x 11.5 x 4 cm). The individual bath and the antechamber were added to this core during the eighteenth century, according to the 76.4 cm arşin then in use (see also [Özdural 1988]). The south-eastern part may have also included an individual bath, with an orthogonal plan and domed cover, which was later destroyed.

The Ottoman *kule*, or defensive tower, in the town of Florina initially had a second storey, of which only part of the eastern façade, comprising a closed opening, the end of the staircase and a second smaller opening, are preserved. The morphology of the tower (pointed Islamic arches with four tracing centres) resemble earlier Ottoman monuments of the sixteenth and seventeenth centuries (see also [Brouskari 2008]). Nevertheless, its construction (stonework with river stones and use of timber ties) suggest a more recent, eighteenth-century construction. Apart from the above, the tracing and the construction of the kule (external dimensions equal to 9 arşin and distance between the timber ties equal to 4 arşin) is based on a 75.8 cm arşin, which was declared as the imperial measuring unit during the years 1794-5 by Selim III (see also [Özdural 1988]). Consequently, its construction can be dated with a fair degree of certainty to the end of the eighteenth or the beginning of the nineteenth century. The octagonal tracing of the walls, arches and squinches is based on the use of a circle with a diameter of 9 arşin, and thus the internal diagonal is equal to the external dimension (fig. 11).

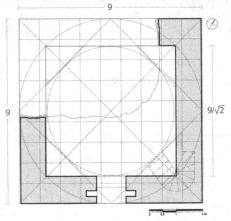

Fig. 11. The Ottoman tower of Florina, plan

At the same time, the internal height is equal to the internal dimensions (9/√2 arşin) and as a result, an imaginary sphere can be inscribed in the interior (fig. 12-13). All the above elements – octagon, double square and sphere – constitute elements of symbolism in Islamic architecture.

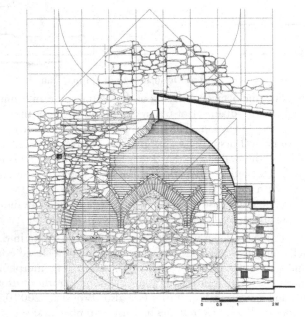

Fig. 12. The Ottoman tower of Florina, section

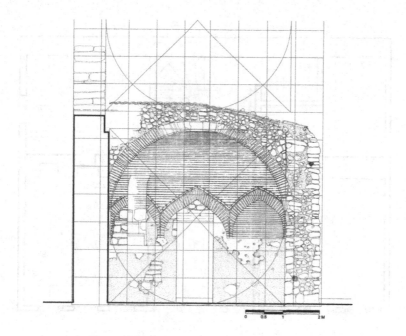

Fig. 13. The Ottoman tower of Florina, section

## 4.2 The traditional houses of Florina

In small buildings like the Mavroudis house (model 5/3/5; see also fig. 2), the dimensions of the odas are 5 x 5 arşin and of the sofa 3 x 5 arşin. The tracing of the ground floor is based on 6-8-10 triangles (3-4-5 multiplied by a factor of 2) (fig. 14). Other buildings (model 5-3), have a plan that is equal to the 2/3 of the model, whereas the duplication of the sofa (model 5-3-3-5), leads to different plans [Oikonomou and Dimitsantou-Kremezi 2009: 311, picture 2].

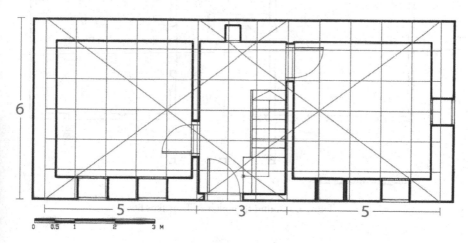

Fig. 14. Ground floor plan of Mavroudis house

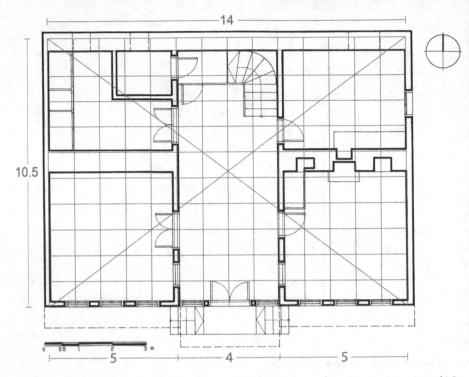

Fig. 15. Ground floor plan of Chatzitzotzas Mansion. Redrawn from [Stoios, Papagiannakis]

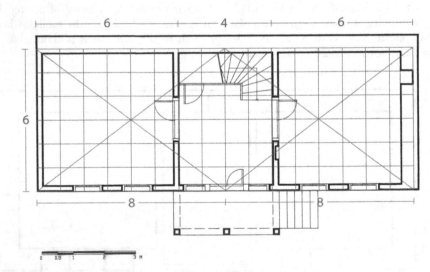

Fig. 16. Ground floor plan of Temelkos house

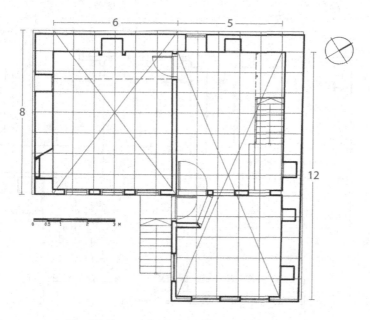

Fig. 17. Ground floor plan of Ioannou Mansion

Chatzitzotzas Mansion (model 5/4/5 - see also fig. 5) is a two-storey house with an inner sofa built around 1875. On the ground floor, the dimensions of the odas are 5×5 arşin and of the secondary rooms are 5×4 arşin. The sofa measures 4×10 arşin. Note that the width of the secondary rooms is equal to that of the sofa. The tracing of the ground floor, including the thick stone wall on the back, is based on a 10.5-14-17.5 triangle (3-4-5 multiplied by a factor of 3.5) (fig. 15).

In Temelkos house (model 6/4/6) the simple tracing uses two 6-8-10 triangles (fig. 16). The odas have a square plan with dimensions 6×6 arşin and the inner sofa measures 4×6 arşin.

In a different house plan deriving from the model 6/5/6, that of Ioannou Mansion, one can see the extension of the inner sofa in the front, a special space known as *divan-hane* with dimensions 5×5 arşin. The tracing is based on the 6-8-10 (3-4-5 multiplied by a factor of 2) and 5-12-13 triangles (fig. 17).

### 4.3 The Mansions of Kastoria

In Kastoria, during the eighteenth century, the building type with a hayat and eyvan, and the same plan (two odas with a hayat and an eyvan) is found in more than one case: the mansions Tsiatsapa (1754, fig. 18) and Sapountzi (fig. 19).[7] The arşin in use is that of the eighteenth century (76.4 cm). The internal dimensions of the odas are 6×6 and the eyvans are 5 arşin wide, whereas the hayat is 8 arşin deep and has an area equal to that of the main spaces. The tracing of the perimeter is made with the use of two 12-16-20 triangles, while two 8-15-17 triangles are used in the tracing of the interior eyvan walls. A very interesting finding is the fact that a circle with a radius of 17 arşin can be used for the tracing of the two kiosks on the façade of the Tsiatsapa Mansion.

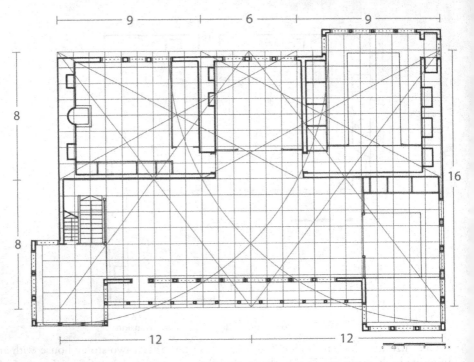

Fig. 18. Upper storey plan of Tsiatsapas Mansion. Plan redrawn from [Mazarakis 2005:a]

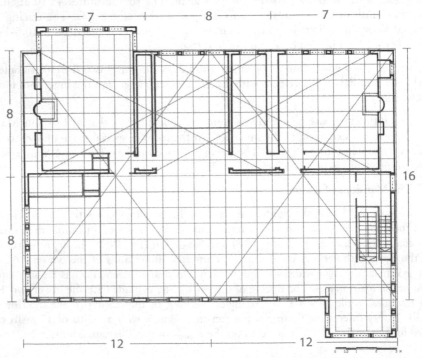

Fig. 19. Upper storey plan of Sapountzis Mansion. Plan redrawn from [Mazarakis 2005:a]

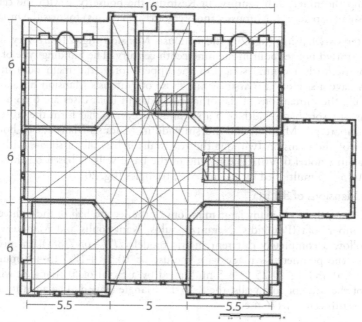

Fig. 20. Upper storey plan of Papaterpos Mansion. Plan redrawn from [Mazarakis 2005:a]

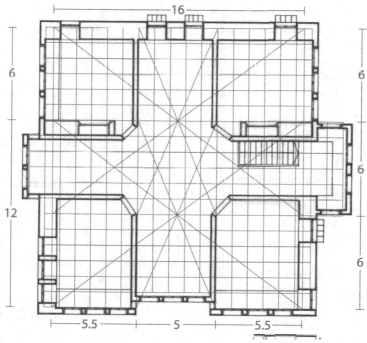

Fig. 21. Upper storey plan of Mitousis Mansion. Plan redrawn from [Papaioannou 2003]

During the nineteenth century, in Kastoria, the house type with the central (cross-shaped) sofa (Papaterpos, Mitoussis and Vergoulas Mansions) follows the model 6/5/6.

Papaterpos Mansion (fig. 20) and Mitousis Mansion (fig. 21) are very similar. Both are three-storied houses built in the nineteenth century. The building type of both is that of the house with a central sofa. On the upper storey, the main odas in Papaterpos Mansion have a size of 5×6 arşin, while those of Mitousis Mansion have a size of 5×7 arşin, while the dimensions of the other two odas in both cases are 6×6 arşin. In both mansions, the sofa has a width of 5 arşin and the eyvans that form the cross, are 5 arşin wide in Papaterpos Mansion and 4 arşin wide in Mitousis Mansion. In both cases, the perimeter of the mansion (without the rear wall) is basically a square with dimensions 17×17 arşin (model 6/5/6). The tracing of the basic walls is based on the 5-12-13 and 12-16-20 (3-4-5 multiplied by a factor of 2) triangles (fig. 20).

## 4.4 The Mansions of Siatista

In Siatista, there are many large mansions of the second half of the eighteenth century with an inner sofa (Poulkidis, Nerantzopoulos, Kanatsoulis and Manousis Mansions), which follow a completely different metric model (7/7/7 or 7/8/7). In most cases, the tracing of the perimeter is made with the use of 10.5-14-17.5 (3-4-5 multiplied by a factor of 3.5) and 15-20-25 (3-4-5 multiplied by a factor of 5) triangles, whereas in the tracing of the interior sofa walls, the 8-15-17 triangle is used. The arşin in use is that of the eighteenth century (76.4 cm).

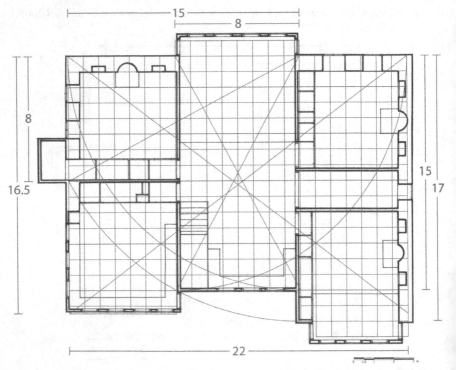

Fig. 22. Upper storey plan of Poulkidis Mansion. Plan redrawn from [Moutsopoulos 1964]

Poulkidis Mansion (1752) and Nerantzopoulos Mansion (1754), probably built by the same master-builders, show an identical tracing of the perimeter based on the triangle 16.5-22-27.5 (3-4-5 multiplied by a factor of 5.5) (fig. 22). The tracing of the interior hall walls is made with the use of the 8-15-17 triangle. A very interesting finding is the fact that these interior walls can be traced with two circles with a radius of 15 arşin (fig. 22) and in that way two squares 15 x 15 are inscribed in the building perimeter. Also, a circle with a radius of 17 arşin can be used for the tracing of characteristic points of wall intersections.

Kanatsoulis Mansion (1757) and Manousis Mansion (1763) show an identical tracing of the perimeter based on the triangle 16.5-22-27.5 (3-4-5 multiplied by a factor of 5.5) (fig. 23). The tracing of the rooms and the interior hall is made with the use of the 8-15-17 triangle. The right interior wall of the hall can be traced with a circle with a radius of 15 arşin (fig. 23) and a second circle with a radius of 17 arşin can be used for the tracing of the main oda projection on the left. The plans of the two Mansions are identical, suggesting that they were built by the same master builders (probably the ones that built Poulkidis and Nerantzopoulos Mansions).

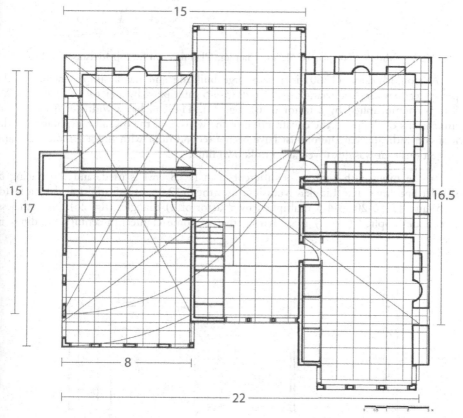

Fig. 23. Upper storey plan of Kanatsoulis Mansion. Plan redrawn from [Mazarakis 2005:b]

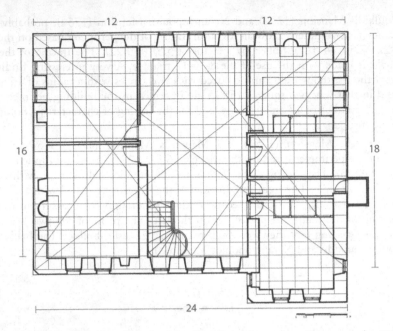

Fig. 24. Middle floor plan of Alexiou Mansion. Plan redrawn from [Moutsopoulos 1964]

Alexiou Mansion (1760) is a very large building and has a simple tracing of the perimeter that uses the triangles 12-16-20 (3-4-5 multiplied by a factor of 4) and 18-24-30 (3-4-5 multiplied by a factor of 6) (fig. 24). The hall has a width of 8 arşin. The dimensions of the plan, the construction of the walls, and the overall morphology of the mansion indicate that there must have been a second floor with timber-frame walls and projections that was destroyed at an unknown time.

In the nineteenth century, a different arsin (75,8 cm) is used, and the tracing of smaller buildings, like Lioukas Mansion, uses the 9-12-15 (3-4-5 multiplied by a factor of 3) and 12-16-20 (3-4-5 multiplied by a factor of 4) triangles (fig. 25). Nevertheless the model applied (7-7-7) is similar to that of some eighteenth-century buildings in Siatista.

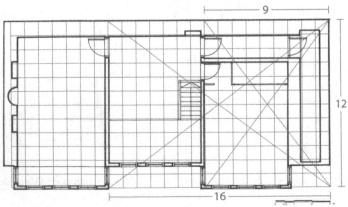

Fig. 25. Upper storey plan of Lioukas Mansion. Plan redrawn from [Moutsopoulos 1964]

### 4.5 The Mansions of Kozani

In Kozani, there are a few eighteenth-century mansions, such as the G. Vourkas Mansion (1748) and the N. Vourkas-Katsikas Mansion, with plans similar to the ones of Siatista. In fact the same plan is applied 'in mirror'. In N. Vourkas-Katsikas Mansion (1762), the tracing of the perimeter is based on the 15-20-25 triangle, while the 8-15-17 triangle is used in the tracing of the interior sofa walls and the rooms (fig. 26).

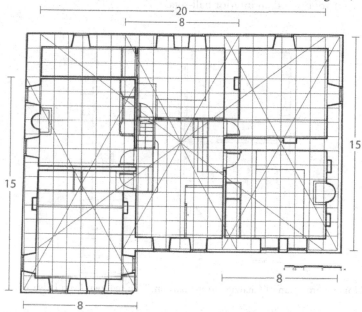

Fig. 26. Upper storey of N. Vourkas Mansion. Plan redrawn from [Haritidou-Mavroudi 1986]

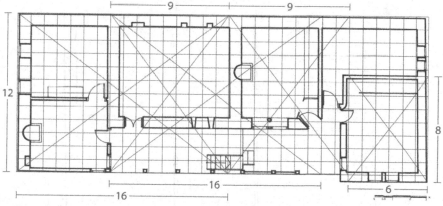

Fig. 27. Upper storey plan of Paloukas Mansion. Plan redrawn from [Mazarakis 2005:b]

In the plan of Paloukas Mansion (fig. 27), the tracing of the perimeter, the hayat and all the interior walls is based entirely on the transformations of the 3-4-5 triangle: 6-8-10, (3-4-5 multiplied by a factor of 2);  9-12-15 (3-4-5 multiplied by a factor of 3); and 12-16-20 (3-4-5 multiplied by a factor of 4).

## 4.6 The Mansions of Eratyra

The Eratyra Mansions have many characteristics in common with these of Siatista and Kozani. In Chatzigiannis Mansion, the tracing of the inner sofa is based on the triangle 8-15-17, and the tracing of the perimeter is made with the use of a circle of a 15 arşin radius (fig. 28). On the ground floor, a combination of a 9-12-15 and a 5-12-13 triangle is used. The applied model is in fact similar to the 2/3 of the completed building type with four rooms and an interior hall-sofa.

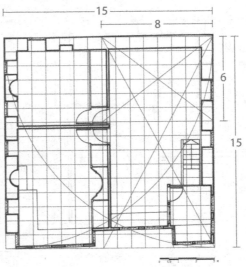

Fig. 28. Upper storey plan of Chatzigiannis Mansion. Plan redrawn from [Aivazoglou-Dova 1998]

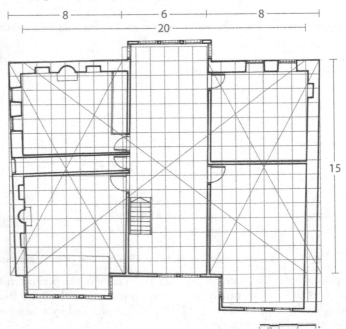

Fig. 29. Upper storey plan of Bairaktarous Mansion. Plan redrawn from [Aivazoglou-Dova 1998]

Lazaridis Mansion (1798) and Bairaktarous Mansion show a similar tracing of the perimeter based on the triangle 15-20-25 (3-4-5 multiplied by a factor of 5) (fig. 29). The tracing of the rooms and the interior hall is made with the use of the 8-15-17 triangle. The plans of the two mansions present many similarities, suggesting that they were built by the same master builders.

## 4.7 The Mansions of Ampelakia

In Ampelakia there existed small houses with an open hayat and with a typology of rural origin, like Tsokanos house. In this building, the simple tracing uses a combination of 9-12-15 and 5-12-13 triangles, and the perimeter is based on a circle with a radius of 13 arşin (fig. 30).

On the other hand, in larger buildings like Krasoulis Mansion (1797), which has an open hayat with an eyvan, the tracing of the perimeter is based on the 13.5-18-22.5 (3-4-5 multiplied by a factor of 4.5) and 9-12-15 triangles. The tracing of the eyvan uses a 5-12-13 triangle (fig. 31) The plan shows similarities with nineteenth-century cross-shaped plans.

Similar tracing can be found in the G. Schwartz Mansion, built in 1798. The L-shaped stone perimeter wall is designed and constructed with the use of an 18-24-30 (3-4-5 multiplied by a factor of 6) triangle, while the tracing of the interior walls is based on the 5-12-13 and 9-12-15 (3-4-5 multiplied by a factor of 3) triangles (fig. 32). All the eyvans are 5 arşins wide.

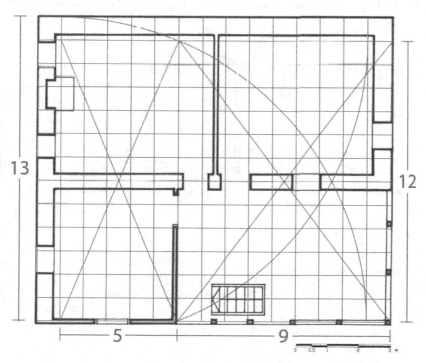

Fig. 30. Upper storey plan of Tsokanos house. Plan redrawn from [Megas 1946]

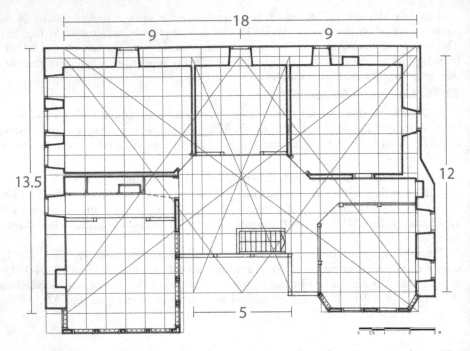

Fig. 31. Upper storey plan of Krasoulis Mansion. Plan redrawn from [Moutsopoulos 1966]

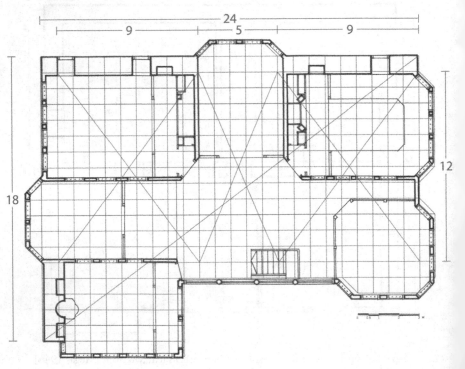

Fig. 32. Upper storey plan of G. Schwartz Mansion. Plan redrawn from [Moutsopoulos 1966]

# 5 Conclusions

In this paper, the application of metric proportions and tracing has been investigated in selected examples of eighteenth- and nineteenth-century mansions and traditional houses in Northern Greece. This has made it possible to identify certain metric models and housing types that were commonly applied, together with the triangles used in their tracing. The basic house types with a central sofa or an inner sofa are, in some cases, adjusted to the size of the plot and the demands of the owner, resulting in smaller houses with two or four odas and an inner sofa with or without eyvan. The three basic metric models are the 5/4/5, the 6/5/6 and 7/7/7.

In Kastoria, during the eighteenth century, the building type with a hayat and eyvan, and the same plan (model 6/5/6 with a hayat 8 arşin deep) is applied to more than one case (Sapountzi and Tsiatsapa Mansions). During the nineteenth century, in Kastoria, the house type with the central sofa (Papaterpos, Mitoussis and Vergoulas Mansions) follows the model 6/5/6.

The same model and the model 5/4/5 are also found in Florina, Trikala, Eratyra, Veria, Xanthi during the nineteenth century, while in towns such as Siatista and Kozani in the second half of the eighteenth century there are mansions with an inner sofa (Nerantzopoulos, Manousis, Poulkidis and Vourkas Mansions) that follow a completely different metric model (7/7/7 or 7/8/7). Another important finding is that the tracing of the Siatista and Kozani Mansions (1740 – 1765), as well as of two Kastoria Mansions (Sapountzi and Tsiatsapa), is based on the arşin of 76.4 cm, whereas in the majority of the other presented examples (1795 and after), the 75.8 cm arşin is used.

Concerning the use of triangular tracing, it should be noted that during the period of the eighteenth and nineteenth centuries, the basic triangle 3-4-5 and its transformations (multiplied by factors of 2, 2.25, 3, 3.5, 4, 4.5, 5, 5.5 and 6) is used for the tracing of the perimeter. Nevertheless, the change in the building types and the space dimensions resulted in a significant change in the use of orthogonal triangles. Consequently, apart from the above-mentioned triangle, during the eighteenth century, the triangle 8-15-17 is widely used in the tracing of hayats and sofas. On the contrary, during the nineteenth century, in the compact house forms, the 5-12-13 triangle is used in the tracing of sofas and eyvans.

From the above, it can be seen that the choice of certain basic models – prototypes and triangles is directly linked to the application of the predominant building types during the different periods and in the different areas (type with hayat and eyvan, type with an inner sofa and type with a central sofa) and largely affects the form and the proportions of the houses.

Further research should include a larger number of mansions from the wider Balkan area, including houses of Albania, countries of the Former Yugoslavia, Bulgaria and Turkey, pertaining to different building types. In that way, the principles and the proportions of the traditional house of the Ottoman Period can be derived and similarities and differences between the different urban centres of the different countries can be noticed.

### Acknowledgments

Aineias Oikonomou would like to thank the Greek State Scholarships Foundation (I.K.Y.) for supporting the post-doctoral research of which this study forms part. The research was supervised by Professor Aikaterini Dimitsantou-Kremezi, and was conducted in the Laboratory of Architectural Form and Orders, N.T.U.A. School of Architecture. The author would also like to thank Arch. Eng. Achilleas Stoios and Arch. Eng., M.Sc., Michalis Papagiannakis for providing the original drawings of the Chatzitzotzas Mansion in Florina.

### Notes

1. Cerasi [1988] cites [Berbenliev and Partasev 1963], as well as his personal communications with P. Berbenliev.
2. Cerasi [1988: 95] also notes that the use of long leather strips for the measuring and tracing on the site encourages circular and diagonal tracing.
3. See also Jouven [1961: 23, 24], for the council concerning the construction of the Milan Cathedral *ad quadratum* or *ad triangulum* (1391), as well as Jouven [1951: 25], for the principle "*circle-square-triangle*" by the Bauhutten building guild in Germany, according to Ghyka [1931]).
4. This investigation departs from the approach described by Cerasi [1988] and Stoichkov [1977].
5. See also Eldem [1955: 223], who mentions that it is very likely that the initial designs on which the basic building types were based were designed by capable Turkish architects.
6. See also Choisy [1976: 49-50] and Konstantinidis [1961: 112] concerning the construction of right angles with the application of the Pythagorean theorem in Egyptian and Indian architecture, respectively. The 3-4-5 triangle is sometimes called the Egyptian Triangle.
7. These two mansions are based on a similar plan, which is applied in a symmetrically opposite way.

### References

AIVAZOGLOU-DOVA, D. 1998. Eratyra. Pp. 275-308 in *Greek Traditional Architecture. Greece vol. 7. Macedonia A'*, G. Lavas, ed. Athens: Melissa Publications (in Greek).

ANGUELOVA, R. 1993. Bulgaria. Pp. 83-148 in *Balkan Traditional Architecture*, L. Vagena Papaioannou and D. Kamini-Dialeti, eds. Athens: Melissa Publications (in Greek).

BERBENLIEV, P. and V. H. PARTASEV. 1963. *Bratzigovskite Maistori Stroiteli prez XVIII i XIX ve i tahnoto architekturno tvorcestvo*. Sofia: Tehnika (in Bulgarian).

BROUSKARI, E. 2008. *Ottoman Architecture in Greece*. Athens: Ministry of Culture Directorate of Byzantine and Post-Byzantine Antiquities (in Greek).

CERASI, M. 1988. Late-Ottoman Architects and Master Builders. Pp. 87-102 in *Muqarnas V: A Annual on Islamic Art and Architecture*, O. Grabar, ed. Leiden: E.J. Brill.

CHOISY, A. 1976. *Histoire de l'Architecture.vol. 1&2* France: Editions SERG.

DIMITRIADIS, V. 1973. *Central and Western Macedonia according to Evliya Celemb* Thessaloniki: Society for Macedonian Studies (in Greek).

ELDEM, S. H. 1955. *Türk Evi Plan Tipleri*. Istanbul: Istanbul Teknik Universitesi – Mimarli Fakultesi, Pulhan Matbaasi.

GHYKA, Matila C. 1931. *Le Nombre d'or*. Paris: NRF.

HARITIDOU-MAVROUDI, E. 1986. To arhontiko tou Nikolaou Vourka (i Elia Katsika) stin Kozai (The mansion of Nikolaos Vourkas (or Nikolaos Katsikas) in Kozani). Pp. 43-56 in *Eponyn Arhontika ton Hronon tis Tourkokratias* (*Eponymous Mansions of the Period of the Turkis Domination*). Athens: National Technical University of Athens.

JOUVEN, G. 1951. *Rythme et Architecture. Les Traces Harmoniques.* Paris: Editions Vincent, Freal et Cie.

KANETAKI, E. I. 2004. *Ottoman Baths in Greece.* Athens: Technical Chamber of Greece Publications (in Greek).

KIZIS, J. 1994. *Pilioritiki Oikodomia (Domestic Architecture in Pelion $17^{th}$ – $19^{th}$ c.).* Athens: Cultural Technological Foundation ETBA (in Greek).

KONSTANTINIDIS, D. 1961. *Peri armonikon haraxeon eis tin arhitektonikin kai tas eikastikas tehnas. Historia kai Aesthitiki (On Harmonic Tracing in Architecture and Visual Arts. History and Aesthetics).* Athens, Greece (in Greek).

KOZUKHAROV, G. 1974. *Svodut v Anticnostta u Srednite Vekove.* Sofia, Bulgaria: Anticnostta (in Bulgarian).

KUBAN, D. 1995. *The Turkish Hayat House.* Istanbul: T.C. Ziraat Bankasi, MTR.

LABORATORY OF THEORY OF FORMS. *Archive.* Xanthi: Section of Architectural Design and Construction, Department of Architecture, Democritus University of Thrace.

LOUKAKIS, P. 1977. *To giannotiko spiti (The house of Giannena).* Pp. 194-228 in *To Elliniko Laiko Spiti (The Greek Popular House),* P. Mihelis, ed. Athens: N.T.U.A (in Greek).

MAZARAKIS-AENIAN, F., ed. 2005. *Arhontika tis Kastorias (Mansions of Kastoria).* Athens: Historical and Ethnological Society of Greece (in Greek).

———, ed. 2005. *Arhontika tis Kozanis kai tis Siatistas (Mansions of Kozani and Siatista).* Athens, Greece: Historical and Ethnological Society of Greece (in Greek).

MEGAS, G. 1946. *Thessalikai oikiseis (Houses in Thessaly).* Athens, Greece: Ministry of Reconstruction Publications (in Greek).

MOUTSOPOULOS, N.K. 1964. Ta arhontika tis Siatistas (The mansions of Siatista). In *Scientific Year-Book of the Polytechnic School, vol. A. 1961-1964.* Thessaloniki: Aristotle University of Thessaloniki (in Greek).

———. 1966. *Ta Thessalika Ampelakia. Isagogi stin istoria, tin koinopraxia kai ta arhontika tis komopoleos (The Thessalian Ampelakia. An introduction to the history, the joint-venture and the mansions of the town).* Athens: Ios Review, "Thessaly" (in Greek).

———. 1967. *I laiki arhitektoniki tis Verias (The popular architecture of Veria).* Athens, Greece: Technical Chamber of Greece Publications (in Greek).

———. 2010. *Naodomia (Construction d'Eglises).* Thessaloniki: University Studio Press (in Greek).

NECIPOĞLU-KAFADAR, G. 1986. Plans and Models in 15th- and 16th-Century Ottoman Architectural Practice. *Journal of the Society of Architectural Historians* **45**, 3: 224-243.

OIKONOMOU, A. 2006. Lessons from traditional architecture on sustainable planning and building design. Pp. 87-94 in *ECOPOLIS: Sustainable Planning and Design Principles. Revealing and Enhancing Sustainable Design,* D. Babalis, ed. Florence: Alinea.

———. 2007. *Comparative Investigation of the Architectural Structure and the Environmental Performance of 19th Century Traditional Houses in Florina.* Ph.D. dissertation. Athens, Greece: National Technical University of Athens, School of Architecture (in Greek).

OIKONOMOU, A. and A. DIMITSANTOU-KREMEZI. 2009. Investigation of the Application of the Module in the Traditional Architecture of Florina. Pp. 309-318 in *Appropriate Interventions for the Safeguarding of Monuments and Historical Buildings. New Design Tendencies. Proceedings of the $3^{rd}$ National Congress,* M. Dousi and P. Nikiforidis, eds. Thessaloniki: Ianos Publications (in Greek).

OIKONOMOU, A., et al. 2009. Application of the Module and Construction Principles in the Traditional Architecture of Northern Greece. Pp. 543-562 in *LIVENARCH IV (RE/DE) Constructions in Architecture. 4th International Congress Liveable Environments and Architecture.* vol. 2. Ş.Ö. Gür, ed. Trabzon, Turkey: Karadeniz Technical University, Faculty of Architecture, Department of Architecture.

OUSTERHOUT, R. 2008. *Master Builders of Byzantium.* Philadelphia: University of Pennsylvania Museum of Archaeology and Anthropology.

ÖZDURAL, A. 1988. Sinan's Arşin: A Survey of Ottoman Architectural Metrology. Pp. 101-115 in *Muqarnas XV: An Annual on the Visual Culture of the Islamic World,* G. Necipoğlu, ed. Leiden: E.J. Brill.

PAPAIOANNOU, K. 2003. *The Traditional Greek House.* Athens: National Technical University of Athens Publications (in Greek).

POPOV, I. 1955. *Proporcii v Bulgarskata Arhitektura (Proportions in Bulgarian Architecture).* Sofia, Bulgaria: Hayka y Yekystvo (in Bulgarian).

SISA, B. 1990. Hungarian Wooden Belltowers of the Carpathian Basin. Pp. 305-352 in *Armos, vol. A.* Thessaloniki, Greece: Aristotle University of Thessaloniki, Polytechnic School, Department of Architecture.

STOICHKOV, J. 1977. Za *Arhitecturata na Koprivchtitza (The Architecture of Koprivchtitza).* Sofia, Bulgaria: Tehnika (in Bulgarian).

## About the author

Aineias Oikonomou is an architect and for two years was an adjunct lecturer at the Departments of Architecture of University of Patras and Democritus University of Thrace. He earned a Ph.D. and M.Sc. degree in Architecture from the National Technical University of Athens, Greece, and he completed a one-year post-doctoral research in the Laboratory of Architectural Form and Orders, N.T.U.A. School of Architecture. His research interests include architectural morphology, restoration, study of pre-industrial building techniques, history of construction technology and Islamic architecture.

Printed in the United States
By Bookmasters